Geometry of Submanifolds

Bang-Yen Chen

Department of Mathematics
Michigan State University

Dover Publications, Inc. | Mineola, New York

Bibliographical Note

This Dover edition, first published in 2019, is an unabridged, corrected republication of the work originally published in 1973 by Marcel Dekker, Inc., New York.

Library of Congress Cataloging-in-Publication Data

Names: Chen, Bang-Yen, author.
Title: Geometry of submanifolds / Bang-Yen Chen.
Description: Dover edition. | Mineola, New York : Dover Publications, Inc.,
  2019. | Originally published: New York : Marcel Dekker, Inc., 1973. |
  Includes bibliographical references and indexes.
Identifiers: LCCN 2018056577 | ISBN 9780486832784 | ISBN 0486832783
Subjects: LCSH: Submanifolds. | Manifolds (Mathematics)
Classification: LCC QA649 .C48 2019 | DDC 516.3/62—dc23
LC record available at https://lccn.loc.gov/2018056577

Manufactured in the United States by LSC Communications
83278301   2019
www.doverpublications.com

Contents

Preface

The theory of submanifolds as a field of differential geometry is as old as differential geometry itself, beginning with the theory of curves and surfaces. However, the theory of submanifolds given in this book is relatively new in the realm of contemporary differential geometry.

The reader is assumed to be somewhat familiar with general theory of differential geometry as can be found, for example, in Kobayashi-Nomizu's *Foundations of Differential Geometry*. Most of the required background material is collected in the first two chapters. In Chapter 1, we have given a brief survey of Riemannian geometry and in Chapter 2, we have given a brief survey of the general theory of submanifolds.

In Chapter 3, minimal submanifolds are studied. Results in this chapter include classical results on the first variation of the volume integral and Bernstein's theorem as well as some recent results of Calabi, do Carmo, Chern, Kobayashi, Lawson, Simons, Takahashi, Wallach, Yano, and the author.

In Chapter 4, submanifolds with parallel mean curvature vector are studied. The theory of analytic functions is applied to the case of surfaces and give a powerful method which was used by Hopf. The main results of this chapter include recent works of Erbacher, Ferus, Hoffman, Ishihara, Klotz, Leung, Ludden, Nomizu, Osserman, Ruh, Smyth, Wolf, Yano, Yau, and the author.

In Chapter 5, conformally flat submanifolds are studied. The results in this chapter were mostly obtained by Yano and the author.

In Chapter 6, submanifolds umbilical with respect to a normal direction are studied. The normal connection of submanifolds play an important role in this chapter. Most results of this chapter were obtained by Nomizu, Smyth, Yano, and the author.

In the last chapter, geometric inequalities of submanifolds are given. Some elementary results in Morse theory are collected in the first section. These results have many important applications to the later sections. Results of Chern and Lashof on total absolute curvature are given in the second section. In sections 3 through 6, the total mean curvature of a submanifolds is studied. The results in these four sections include recent work of Shiohama, Takagi, Willmore, and the author. In the last section, stable hypersurfaces with respect to the total mean curvature are studied.

At the end of each chapter, some problems are given. These problems can be regarded as supplements to the text.

Since this book is based primarily on the author's recent work on real submanifolds, omissions of important results are inevitable.

In concluding the preface, the author would like to thank Professor S.S. Chern and Professor S. Kobayashi, who invited the author to undertake this project. The author also likes to thank Professor S.S. Chern, Professor T. Nagano, and Professor T. Ōtsuki for their constant encouragement and guidance. Finally, the author is greatly indebted to his colleagues, Dr. D.E. Blair, Dr. G.D. Ludden, and Dr. K. Ogiue for their help, which resulted in many improvements of both of the content and the presentation. Finally, the author wishes to thank Mrs. Mary Reynolds, who typed the manuscript, for her patience and cooperation.

Bang-Yen Chen

Chapter 1

Riemannian Manifolds

1 Riemannian Manifolds

We consider an n-dimensional manifold* M of class C^∞ covered by a system of coordinate neighborhoods $\{U, x^h\}$, where U denotes a neighborhood and the x^h denote local coordinates in U, with the indices h, i, j, k, l taking on values in the range $1, 2, \ldots, n$.

If, for any system of coordinate neighborhoods covering the manifold M, there exist a finite number of the coordinate neighborhoods which cover the whole manifold, then the manifold M is said to be *compact*.

If we can cover the whole manifold M by a system of coordinate neighborhoods in such a way that the Jacobian determinant

$$|\partial x^{h'}/\partial x^h| \tag{1.1}$$

of the coordinate transformation

$$x^{h'} = x^{h'}(x^1, \ldots, x^n) \tag{1.2}$$

in every nonempty intersection of two coordinate neighborhoods $\{U; x^h\}$ and $\{U'; x^{h'}\}$ is always positive, then the manifold M is said to be *orientable*.

---

*We consider only connected manifolds of dimension > 1 unless otherwise stated.

We assume that there is given on M a positive-definite symmetric tensor field g of type (0,2) and of differentiability class C^∞. We call such a manifold M a *Riemannian manifold*.

In the following, we denote by

$$\partial_i = \partial/\partial x^i$$

the basis vectors on the coordinate neighborhood $\{U; x^h\}$. Let X and Y be two vector fields on M. We then have

$$X = X^h \partial_h, \quad Y = Y^h \partial_h,$$

where X^h and Y^h are the local components of the vector fields X and Y, respectively, with respect to the natural frame ∂_h, where, here and in the sequel, we shall use the Einstein convention, that is, repeated indices, with one upper index and one lower index, denoted summation over its range.

We then have

$$g(X, Y) = g(X^j \partial_j, Y^i \partial_i) = g_{ji} X^j Y^i, \tag{1.3}$$

where

$$g_{ji} = g(\partial_j, \partial_i) \tag{1.4}$$

are the local components of the metric tensor field g.

For a differentiable curve $\sigma(t)$;

$$x^h = x^h(t), \quad a \leqq t \leqq b,$$

we define the arc length s by the definite integral

$$s = \int_a^b \sqrt{g_{ji}(x(t)) \frac{dx^j}{dt} \frac{dx^i}{dt}} \, dt.$$

Since we have a metric tensor g on M, we can define the length of a vector X by

$$|X| = \sqrt{g(X, X)}, \tag{1.5}$$

and the angle θ between two vectors X and Y at the same point by

$$\cos \theta = \frac{g(X, Y)}{|X||Y|}. \tag{1.6}$$

Since the quadratic form $g(X, X)$ is positive definite, the determinant

$$\mathfrak{g} = |g_{ji}|$$

formed with local components g_{ji} of g is always positive, and consequently, we can find g^{ih} such that

$$g_{ji} g^{ih} = \delta_j^h, \tag{1.7}$$

where δ_j^h is the Kronecker delta, that is,

$$\delta_j^h = 1, \quad h = j$$
$$\quad = 0, \quad h \neq j.$$

We call g^{ih} the *contravariant components* of the metric tensor g and g_{ji} the *covariant components*.

We use the covariant components g_{ji} and contravariant components g^{ih} to lower and to raise the indices of the components of a tensor, for example,

$$T_{jih} = T_{ji}^t g_{th}$$

and

$$T_{ji}^h = T_{jit} g^{th}.$$

We put

$$|T|^2 = T_{jih} T^{jih} \tag{1.8}$$

and call $|T|$ the length of T. Generalizing this, we also use the notation

$$g(S, T) = S_{jih} T^{jih}$$

for two tensor fields S and T of the same type with local components S_{jih} and T_{jih}, respectively.

When the Riemannian manifold M is orientable, we can define the volume element of M by

$$dV = \sqrt{\mathfrak{g}}\, dx^1 \wedge dx^2 \wedge \cdots \wedge dx^n, \tag{1.9}$$

which is always positive, where $\wedge$ denotes the exterior product, and we can consider the integral of a scalar function f,

$$\int_D f(x) dV,$$

over a domain D of M.

In the following, we shall denote by $T_P(M)$ the tangent space of a manifold M at a point $P \in M$ and by $T(M)$ the tangent bundle of M.

2   Covariant Differentiation

We now construct the Christoffel symbols

$$\begin{Bmatrix} h \\ ji \end{Bmatrix} = \frac{g^{th}}{2}(\partial_j g_{it} + \partial_i g_{jt} - \partial_t g_{ji}). \tag{2.1}$$

We denote by ∇_X the operator of covariant differentiation along the vector field X with respect to the Christoffel symbols. Hence, for a scalar f, we have the covariant derivative $\nabla_X f$ of f along the vector field X. $\nabla_X f$ has local components

$$\nabla_X f \colon Xf = X^i \partial_i f, \tag{2.2}$$

where X^i are the local components of X.

For a vector field Y we have the covariant derivative $\nabla_X Y$ of Y along the vector field X. $\nabla_X Y$ has local components

$$\nabla_X Y \colon X^j \nabla_j Y^h = X^j \left(\partial_j Y^h + \left\{ \begin{matrix} h \\ ji \end{matrix} \right\} Y^i \right), \tag{2.3}$$

where X^h and Y^h are the local components of X and Y, respectively.

For a 1-form ω, we have the covariant derivative $\nabla_X \omega$ of ω along X, which is a 1-form defined by

$$(\nabla_X \omega)(Y) = \nabla_X(\omega(Y)) - \omega(\nabla_X Y) \tag{2.4}$$

for any vector fields X and Y. $\nabla_X \omega$ has local components

$$\nabla_X \omega \colon X^j \nabla_j \omega_i = X^j \left(\partial_j \omega_i - \left\{ \begin{matrix} h \\ ji \end{matrix} \right\} \omega_h \right), \tag{2.5}$$

where ω_i and X^j are the local components of ω and X, respectively.

For a general tensor field, say T, of type (1,2), we have the covariant derivative $\nabla_X T$ along the vector field X, which is a tensor field of the same type and is defined by

$$(\nabla_X T)(Y, Z) = \nabla_X(T(Y, Z)) - T(\nabla_X Y, Z) - T(Y, \nabla_X Z) \tag{2.6}$$

for any vector fields X, Y, and Z. $\nabla_X T$ has local components

$$\nabla_X T \colon X^k \nabla_k T^h_{ji} = X^k \left(\partial_k T^h_{ji} + \left\{ \begin{matrix} h \\ kt \end{matrix} \right\} T^t_{ji} - \left\{ \begin{matrix} t \\ kj \end{matrix} \right\} T^h_{ti} - \left\{ \begin{matrix} t \\ ki \end{matrix} \right\} T^h_{jt} \right), \tag{2.7}$$

where T^h_{ji} and X^k are the local components of T and X, respectively.

In general, for a tensor field S of type (p, q) the covariant derivative $\nabla_X S$ of S along X defines a tensor field of type $(p, q + 1)$, which is denoted by ∇S. For example, for a tensor field, say T, of type (1,2), ∇T is defined by

$$(\nabla T)(X, Y, Z) = (\nabla_X T)(Y, Z). \tag{2.8}$$

∇T has local components

$$\nabla T \colon \nabla_k T^h_{ji} = \partial_k T^h_{ji} + \left\{ \begin{matrix} h \\ kt \end{matrix} \right\} T^t_{ji} - \left\{ \begin{matrix} t \\ kj \end{matrix} \right\} T^h_{ti} - \left\{ \begin{matrix} t \\ ki \end{matrix} \right\} T^h_{jt}, \tag{2.9}$$

where T^h_{ji} are the local components of T.

In particular, for a scalar f, $\nabla_X f$ defines a 1-form ∇f. ∇f is sometimes denoted by df.

We note here that

$$\nabla_X Y - \nabla_Y X - [X, Y] = 0 \qquad (2.10)$$

for any vector fields X, Y, which means that the connection ∇ has no torsion, where $[X, Y]$ represents the vector field defined by

$$[X, Y]f = X(Yf) - Y(Xf) \qquad (2.11)$$

for an arbitrary function f.

We can also define the covariant derivative of relative tensor fields. For example, the covariant derivative of a relative scalar f of weight p is given by

$$\nabla_i f = \partial_i f - p \begin{Bmatrix} t \\ it \end{Bmatrix} f. \qquad (2.12)$$

It is well known that the covariant derivatives of g_{ji}, g^{ih}, δ_i^h, $\sqrt{g}$ all vanish:

$$\nabla_k g_{ji} = \partial_k g_{ji} - \begin{Bmatrix} t \\ kj \end{Bmatrix} g_{ti} - \begin{Bmatrix} t \\ ki \end{Bmatrix} g_{jt} = 0, \qquad (2.13)$$

$$\nabla_k g^{jh} = \partial_k g^{jh} + \begin{Bmatrix} j \\ kt \end{Bmatrix} g^{th} + \begin{Bmatrix} h \\ kt \end{Bmatrix} g^{jt} = 0, \qquad (2.14)$$

$$\nabla_i g = \partial_i g - 2 \begin{Bmatrix} t \\ it \end{Bmatrix} g = 0, \qquad (2.15)$$

$$\nabla_j \delta_i^h = \partial_j \delta_i^h + \begin{Bmatrix} h \\ jt \end{Bmatrix} \delta_i^t - \begin{Bmatrix} t \\ ji \end{Bmatrix} \delta_t^h = 0. \qquad (2.16)$$

Let $\sigma(t)$: $x^i = x^i(t)$ be a curve in the Riemannian manifold M. Then the *tangent vector* of the curve $\sigma(t)$ is given by

$$T(\sigma) = T^i(\sigma)\partial_i, \quad T^i(\sigma) = \frac{dx^i}{dt}.$$

If we have $\nabla_T T = 0$ identically on the curve σ, that is,

$$\frac{d^2 x^h}{dt^2} + \begin{Bmatrix} h \\ ji \end{Bmatrix} \frac{dx^j}{dt} \frac{dx^i}{dt} = 0,$$

then the curve σ is called a *geodesic* in M. From the fundamental theorem of ordinary differential equations, we know that for any vector X at a point $P \in M$, there exists locally a unique geodesic in M passing through P with the initial conditions; $\sigma(0) = P$ and $T(P) = X$. For each vector X at P, let $\sigma(t)$: $x^i = x^i(t)$ be the geodesic, with the initial conditions $(\sigma(0), T(P)) = (P, X)$. We set

$$\exp_P X = \sigma(1).$$

Then we have a mapping of $T_P(M)$ into M for each point P. We call this mapping the *exponential map*. It is well-known that, for each $P \in M$, the exponential map $\exp_P$ at P is a diffeomorphism of a neighborhood N_P of 0 in $T_P(M)$ onto a neighborhood U_P of P in M. Let $E_1, \ldots, E_n$ be an orthonormal basis of $T_P(M)$ at P. Then every vector $X \in T_P(M)$ has some coordinates with X^i with $X = X^i E_i$. Therefore, the diffeomorphism $\exp_P$: $N_P \to U_P$ defines a local coordinate system in U_P in a natural manner. We call this coordinate system on U_P the *normal coordinate system* in U_P. It can be verified that if $\left\{ {h \atop ji} \right\}$ are the Christoffel symbols for this coordinate system, then $\left\{ {h \atop ji} \right\} = 0$ at the point P. Moreover, for constants $a^1, \ldots, a^n$, the curve given by $x^i = a^i t$ is a geodesic emanating from P at $t = 0$.

Let X be a vector at $P \in M$ and $\sigma(t)$ be the geodesic with initial conditions $\sigma(0) = P$ and $T(P) = X$. Then the vector $T(\sigma(t))$ is called the *parallel translate* of X along the geodesic σ.

We now consider a p-form

$$\omega = \frac{1}{p!} \omega_{i_1 i_2 \cdots i_p}\, dx^{i_1} \wedge dx^{i_2} \wedge \cdots \wedge dx^{i_p}$$

or a skew-symmetric tensor of type $(0, p)$. The *exterior differential* or simply *differential* $d\omega$ of ω is the $(p+1)$-form, defined by

$$d\omega = \frac{1}{(p+1)!} (\partial_i \omega_{i_1 i_2 \cdots i_p} - \partial_{i_1} \omega_{i i_2 \cdots i_p} - \partial_{i_2} \omega_{i_1 i i_3 \cdots i_p} - \cdots - \partial_{i_p} \omega_{i_1 i_2 \cdots i})$$
$$\times dx^i \wedge dx^{i_1} \wedge \cdots \wedge dx^{i_p}. \tag{2.17}$$

It is easy to verify that, for any p-form ω,

$$d^2\omega = d(d\omega) = 0. \tag{2.18}$$

For any p-form ω, the *codifferential* $\delta\omega$ of ω is the $(p-1)$-form, with the local expression

$$\delta\omega = -\frac{1}{(p-1)!} (g^{ji} \nabla_j \omega_{i i_2 \cdots i_p}) dx^{i_2} \wedge \cdots \wedge dx^{i_p}. \tag{2.19}$$

If f is a scalar, we put $\delta f = 0$. It can be verified that

$$\delta^2\omega = \delta(\delta\omega) = 0 \tag{2.20}$$

for any p-form ω.

We can also define the codifferential δT of a more general tensor field T. For example, let T be a tensor field of type (0,3). Then δT is the tensor field of type (0,2):

$$\delta T: \ -g^{tj}\nabla_t T_{jih} = -\nabla^j T_{jih}, \tag{2.21}$$

where $\nabla^j = g^{tj}\nabla_t$ and T_{jih} are the local components of T.

A p-form ω, or a skew-symmetric tensor field of type $(0,p)$, is said to be *harmonic* if we have

$$d\omega = \delta\omega = 0. \tag{2.22}$$

If we put

$$\Delta = -\delta d - d\delta, \tag{2.23}$$

then we have $\Delta\omega = 0$ for a harmonic p-form ω.

For a vector field X with local components X^h there is associated a 1-form ξ given by

$$\xi = g_{ji}X^j dx^i = X_i dx^i.$$

The codifferential $\delta\xi$ of ξ is given by

$$\delta\xi = -\nabla_i X^i = -g^{ji}\nabla_j X_i. \tag{2.24}$$

We denote it by δX. The famous Green's theorem can now be stated as follows.

Green's Theorem. *Let X be a vector field on an oriented closed Riemannian manifold M. Then we have*

$$\int_M (\delta X)dV = 0, \tag{2.25}$$

where a closed manifold means a compact manifold without boundary.

For a scalar function f, if we take the codifferential of its covariant derivative $\nabla_i f$. Then we obtain

$$\Delta f = g^{ji}\nabla_j \nabla_i f = \nabla^i \nabla_i f. \tag{2.26}$$

The differential operator $\Delta = g^{ji}\nabla_j\nabla_i$ or $\nabla^i\nabla_i$ is sometimes called the *Laplacian*.

Applying Green's theorem, we have

Theorem 2.1. *For any function f on an orientable closed Riemannian manifold M, we have*

$$\int_M \Delta f dV = 0. \tag{2.27}$$

Using this theorem we have

Hopf's Lemma. *Let M be a closed Riemannian manifold. If f is a function on M such that $\Delta f \geq 0$ everywhere (or $\Delta f \leq 0$ everywhere), then f is a constant function.*

Proof. We may assume that M is orientable by taking the twofold covering of M if necessary. Then, by Theorem 2.1, we have $\Delta f = 0$ everywhere on M. Since

$$\Delta f^2 = 2f\Delta f + g^{ji}(\nabla_j f)(\nabla_i f), \tag{2.28}$$

by using Theorem 2.1 for f^2 and the fact that $\Delta f = 0$ as we have shown, we obtain

$$\int_M g^{ji}(\nabla_j f)(\nabla_i f)dV = 0.$$

This implies $\nabla_j f = 0$ everywhere, that is, f is a constant function. This proves the lemma. □

Let M be a surface, that is, a 2-dimensional manifold, with a Riemannian metric g, and let Δ be the Laplacian on M formed with g. A function f on M is called a *subharmonic* (resp. *superharmonic*) function on M if we have $\Delta f \geq 0$ (resp. $\Delta f \leq 0$) everywhere. A surface M is said to be *parabolic* if there exists no nonconstant negative subharmonic function on M. Thus, if M is parabolic then every subharmonic function on M which is bounded from above on M must be a constant function on M. It is well-known that the euclidean plane E^2 with its standard metric is parabolic.

3   Curvature Tensor

It is known that the connection ∇ defined by the Christoffel symbols is the unique connection which has no torsion and satisfies $\nabla_k g_{ji} = \nabla_k g^{jh} = 0$. We call this connection the *Riemannian connection*. In the following, we consider the Riemannian connection.

Let X, Y, and Z be three vector fields in M. Then

$$\nabla_X \nabla_Y Z - \nabla_Y \nabla_X Z - \nabla_{[X,Y]} Z$$

defines a tensor field of type (1,3). We put

$$\nabla_X \nabla_Y Z - \nabla_Y \nabla_X Z - \nabla_{[X,Y]} Z = K(X,Y)Z. \tag{3.1}$$

Then $K(X, Y)$ is a tensor field of type (1,1) which is linear in X and Y.

With respect to local components, (3.1) can be written as

$$\nabla_k \nabla_j Z^h - \nabla_j \nabla_k Z^h = K^h_{kji} Z^i, \tag{3.2}$$

where

$$K^h_{kji} = \partial_k \begin{Bmatrix} h \\ ji \end{Bmatrix} - \partial_j \begin{Bmatrix} h \\ ki \end{Bmatrix} + \begin{Bmatrix} h \\ kt \end{Bmatrix} \begin{Bmatrix} t \\ ji \end{Bmatrix} - \begin{Bmatrix} h \\ jt \end{Bmatrix} \begin{Bmatrix} t \\ ki \end{Bmatrix} \tag{3.3}$$

are the local components of K. We call this tensor the *Riemann-Christoffel curvature tensor* of the Riemannian manifold M. It gives all the local properties of the Riemannian manifold M.

If we take a 1-form ω, then we have

$$(\nabla_X \nabla_Y \omega)(Z) - (\nabla_Y \nabla_X \omega)(Z) - (\nabla_{[X,Y]} \omega)(Z) = -\omega(K(X, Y), Z). \qquad (3.4)$$

With respect to local components, (3.4) can be written as

$$\nabla_k \nabla_j \omega_i - \nabla_j \nabla_k \omega_i = -K_{kji}^h \omega_h. \qquad (3.5)$$

If we take a general tensor, say T of type (1,2), we have

$$(\nabla_X \nabla_Y T)(Z, U) - (\nabla_Y \nabla_X T)(Z, U) - (\nabla_{[X,Y]} T)(Z, U)$$
$$= K(X, Y)T(Z, U) - T(K(X, Y)Z, U) - T(Z, K(X, Y)U), \qquad (3.6)$$

or in local components

$$\nabla_l \nabla_k T_{ji}^h - \nabla_k \nabla_l T_{ji}^h = K_{lkt}^h T_{ji}^t - K_{lkj}^t T_{ti}^h - K_{lki}^t T_{jt}^h, \qquad (3.7)$$

where T_{ji}^h are the local components of T and X, Y, Z, and U are arbitrary vector fields.

Formulas (3.1), (3.2), (3.4), (3.5), (3.6), and (3.7) are called *Ricci identities*. If $E_1, E_2, \ldots, E_n$ are local orthonormal vector fields, then

$$R(Y, Z) = \sum_{i=1}^{n} g(K(E_i, Y)Z, E_i) \qquad (3.8)$$

defines a global tensor field R of type (0,2) with local components

$$K_{ji} = K_{tji}^t = g^{ts} K_{tjis}. \qquad (3.9)$$

Moreover, from the tensor field R we can define a global scalar field

$$r = \sum_{i=1}^{n} R(E_i, E_i), \qquad (3.10)$$

with local components

$$r = g^{ji} K_{ji}. \qquad (3.11)$$

The tensor field R and the function r are called the *Ricci tensor* and the *scalar curvature*, respectively. If $n = 2$, then $G = \frac{1}{2}r$ is called the *Gaussian curvature*.

From the definition of the curvature tensor K, it is easily seen that K satisfies the following algebraic identities:

$$K(X, Y) + K(Y, X) = 0, \tag{3.12}$$

$$K(X, Y)Z + K(Y, Z)X + K(Z, X)Y = 0, \tag{3.13}$$

or in local components

$$K_{kji}^h + K_{jki}^h = 0, \tag{3.14}$$

$$K_{kji}^h + K_{jik}^h + K_{ikj}^h = 0. \tag{3.15}$$

Consequently, if we put

$$K(X, Y; U, V) = g(K(X, Y)U, V), \tag{3.16}$$

or in local components

$$K_{kjih} = K_{kji}^t g_{th}, \tag{3.17}$$

then we have

$$K(X, Y; U, V) + K(Y, X; U, V) = 0, \tag{3.18}$$

$$K(X, Y; U, V) + K(Y, U; X, V) + K(U, X; Y, V) = 0, \tag{3.19}$$

or in local components

$$K_{kjih} + K_{jkih} = 0, \tag{3.20}$$

$$K_{kjih} + K_{jikh} + K_{ikjh} = 0. \tag{3.21}$$

Equations (3.13), (3.15), (3.19), and (3.21) are called the *first Bianchi identities*.

Applying the Ricci identities to the metric tensor g, we have

$$0 = (\nabla_X \nabla_Y g)(Z, U) - (\nabla_Y \nabla_X g)(Z, U) - (\nabla_{[X,Y]} g)(Z, U)$$
$$= -K(X, Y; Z, U) - K(X, Y; U, Z), \tag{3.22}$$

from which

$$K(X, Y; Z, U) + K(X, Y; U, Z) = 0, \tag{3.23}$$

or in local components

$$K_{kjih} + K_{kjhi} = 0. \tag{3.24}$$

From (3.18), (3.19), and (3.23), we have

$$K(X, Y; Z, U) = K(Z, U; X, Y), \tag{3.25}$$

or in local components

$$K_{kjih} = K_{ihkj}. \tag{3.26}$$

Using the relations satisfied by the curvature tensor K, we can easily prove that the Ricci tensor R is a symmetric tensor, that is,

$$R(Y, Z) = R(Z, Y), \qquad (3.27)$$

or in local components

$$K_{ji} = K_{ij}. \qquad (3.28)$$

For the covariant derivative of the curvature tensor K, we can prove that

$$(\nabla_X K)(Y, Z) + (\nabla_Y K)(Z, X) + (\nabla_Z K)(X, Y) = 0, \qquad (3.29)$$

or in local components

$$\nabla_l K^h_{kji} + \nabla_k K^h_{jli} + \nabla_j K^h_{lki} = 0, \qquad (3.30)$$

$$\nabla_l K_{kjih} + \nabla_k K_{jlih} + \nabla_j K_{lkih} = 0, \qquad (3.31)$$

which is called the *second Bianchi identity*.

Let X and Y be two linearly independent vectors at a point P and $\gamma(X, Y)$ be the plane section spanned by X and Y. The sectional curvature $k(\gamma)$ for γ is defined by

$$k(\gamma) = -\frac{K(X, Y; X, Y)}{g(X, X)g(Y, Y) - g(X, Y)^2}. \qquad (3.32)$$

It is easy to see that this $k(\gamma)$ is uniquely determined by the plane section γ and is independent of the choice of X and Y on it.

If $k(\gamma)$ is a constant for all plane sections γ in the tangent space $T_P(M)$ at P and for all points $P \in M$, then M is called a *space of constant curvature*.

The following theorem is well known.

Theorem 3.1 (Schur, 1886). *Let M be a Riemannian manifold of dimension $n > 2$. If the sectional curvature $K(\gamma)$ depends only on the point P, then M is a space of constant curvature.*

Proof. Let S be the tensor field of type (0,4) defined by

$$S(Z, U; X, Y) = g(Z, X)g(Y, U) - g(U, X)g(Y, Z),$$

where X, Y, Z, U are arbitrary vector fields in M. Then we have

$$K = kS, \qquad (3.33)$$

where k is a function on M. Since $\nabla g = 0$, we have $\nabla S = 0$. Hence, we have

$$(\nabla_W K)(Z, U; X, Y) = (\nabla_W k)S(Z, U; X, Y),$$

for any vector field W, from which

$$((\nabla_W K)(X, Y))Z = (Wk)(g(Z, Y)X - g(Z, X)Y). \qquad (3.34)$$

Taking the cyclic sum of (3.34) with respect to W, X, Y and applying the second Bianchi identity (3.29), we find

$$(Wk)(g(Z, Y)X - g(Z, X)Y) + (Xk)(g(Z, W)Y - g(Z, Y)W)$$
$$+ (Yk)(g(Z, X)W - g(Z, W)X) = 0. \tag{3.35}$$

In particular, if we choose X, Y, Z, and W in such a way that X, Y, $Z = W$ are mutually orthogonal with $g(Z, Z) = 1$. Then we find

$$(Xk)Y - (Yk)X = 0.$$

From this we obtain $Xk = Yk = 0$. Since X can be chosen as an arbitrary vector field, we see that the function k is a constant. This proves the theorem.   □

From Eq. (3.33) we have

$$K(X, Y)Z = k(g(Y, Z)X - g(X, Z)Y) \tag{3.36}$$

for a space of constant curvature k. This equation can also be written as

$$K_{kji}^{h} = \frac{r}{n(n-1)} (\delta_k^h g_{ji} - \delta_j^h g_{ki}), \tag{3.37}$$

where r is the scalar curvature.

From the definition of Christoffel symbols, we can prove that, under a coordinate transformation

$$x'^h = x'^h(x^1, \ldots, x^n),$$

the Christoffel symbols satisfy the following transformation law:

$$\frac{\partial^2 x'^h}{\partial x^j \partial x^i} = \frac{\partial x'^h}{\partial x^t} \left\{ \begin{matrix} t \\ ji \end{matrix} \right\} - \frac{\partial x'^t}{\partial x^j} \frac{\partial x'^s}{\partial x^i} \left\{ \begin{matrix} h \\ ts \end{matrix} \right\}', \tag{3.38}$$

where $\left\{ \begin{matrix} h \\ ts \end{matrix} \right\}'$ denote the Christoffel symbols with respect to the coordinate system $\{x'^h\}$.

If a Riemannian manifold M is of zero curvature, then we have

$$K_{kji}^h = 0. \tag{3.39}$$

In this case, the equations

$$\frac{\partial^2 x'^h}{\partial x^j \partial x^i} = \frac{\partial x'^h}{\partial x^t} \left\{ \begin{matrix} t \\ ji \end{matrix} \right\}$$

obtained from (3.38) by putting $\left\{ \begin{matrix} h \\ ts \end{matrix} \right\}' = 0$ are completely integrable, and therefore there exists a coordinate system in which $\left\{ \begin{matrix} h \\ ts \end{matrix} \right\}' = 0$ and consequently, $g'_{jk} = $ constant.

Conversely, if there exists in M a system of coordinate neighborhoods such that the components of the metric tensor are all constant in any coordinate neighborhood, then we have $\left\{ \begin{matrix} h \\ ji \end{matrix} \right\} = 0$ and consequently, the curvature tensor K vanishes, that is, M is a space of zero curvature. We call such a Reimannian manifold a *locally flat space* or a *locally euclidean space*.

Now, we take a vector X at a point $P \in M$ and consider $n-1$ vectors $X_2, \ldots, X_n$ at P which are orthogonal to X and also mutually orthogonal. Then

$$\lambda = \frac{R(X, X)}{g(X, X)} \tag{3.40}$$

is the sum of sectional curvatures at P with respect to the plane sections given by $(X, X_2), \ldots, (X, X_n)$, and it is independent of the choice of the $n-1$ vectors $X_2, \ldots, X_n$. We call the number λ the *Ricci curvature* at P with respect to the vector X.

If the Ricci curvature at P is independent of the vector X, then the Ricci tensor must be of the form

$$R = \lambda g.$$

If this is the case at every point of the manifold M, then the Ricci tensor has the above form everywhere. If $n > 2$, then, by the second Bianchi identity, we see that λ is a constant. If $n = 2$, the Ricci tensor always has the above form and λ is not necessarily a constant. A Riemannian manifold whose Ricci tensor is of the above form is called an *Einstein space*.

A Riemannian manifold is called a *locally symmetric space* if its curvature tensor is covariant constant, that is $\nabla K = 0$. A complete locally symmetric space is called a *symmetric space*. (For the definition of complete Riemannian manifolds, see the next section). It is clear that every Riemannian manifold of constant curvature is locally symmetric.

4 Space Forms

Suppose that for any two points P_1 and P_2 in a Riemannian manifold M, we define $d(P_1, P_2)$ as the greatest lower bound of the lengths of all piecewise differentiable curves joining P_1 and P_2. Then it can be verified that d defines a metric on M. If d is complete, that is, all Cauchy sequences converge, we say that the Riemannian manifold with the Riemannian metric g is *complete*. It is well-known that any two points P_1 and P_2 in M can be joined by a geodesic arc whose length is equal to the distance $d(P_1, P_2)$ (Hopf and Rinow, 1931) . It is also well-known that the following conditions on M are equivalent:

(a) M is complete,

(b) All bounded closed sets are compact,

(c) Any geodesic arc can be extended in both directions indefinitely with respect to the arc length.

It is clear that every closed Riemannian manifold is complete. Let M be a Riemannian manifold with metric tensor g and M' be another Riemannian manifold with metric tensor g'. If there exists a one-to-one differentiable mapping from M onto M' such that the length of any arc in M measured by g is always equal to the length of the corresponding arc in M' measured by g', then the Riemannian manifolds (M, g) and (M', g') are said to be *isometric*. If there exists a one-to-one differentiable mapping from M into M' such that the angle of any two vectors at a point of M is always equal to that of the corresponding two vectors at the corresponding point of M', then the Riemannian manifolds (M, g) and (M', g') are said to be *conformal*.

It is well-known that any two simply connected complete Riemannian manifolds of constant curvature k are isometric to each other (see, for instance, Kobayashi and Nomizu, 1969). A complete simply connected Riemannian manifold of constant curvature is called a *space form*.* A space form is said to be *elliptic*, *hyperbolic*, or *euclidean* according as the sectional curvature is positive, negative, or zero.

In the following, we shall construct, for each constant k, a space form with curvature k.

Let E^n be the affine n-space with natural coordinates $x^1, \ldots, x^n$ and let g be the euclidean metric on E^n, that is,

$$g = (dx^1)^2 + \cdots + (dx^n)^2.$$

Then E^n with the euclidean metric g forms a space form of zero curvature. We call it the *euclidean n-space* and we denote it by E^n.

We put

$$R^n(k) = \{(x^1, \ldots, x^{n+1}) \in E^{n+1} : \sqrt{|k|}((x^1)^2 + \cdots + (x^n)^2 + \text{sgn}(k)(x^{n+1})^2)$$
$$- 2x^{n+1} = 0, x^{n+1} \geq 0\}, \tag{4.1}$$

where $\text{sgn}(k) = 1$ or -1 according as $k \geq 0$ or $k < 0$. Then the Riemannian connection induced by

$$g = (dx^1)^2 + \cdots + (dx^n)^2 + \text{sgn}(k)(dx^{n+1})^2 \tag{4.2}$$

on E^{n+1} is the ordinary euclidean connection for each value of k. In each case the metric tensor induced on $R^n(k)$ is complete and of constant curvature k. Moreover, each $R^n(k)$ is simply connected. Hence, each $R^n(k)$ with the metric tensor (4.2) gives a model of a space form of curvature k.

---

*In some books, a space form is defined as a Riemannian manifold of constant curvature.

The *hyperspheres* in $R^n(k)$ are those hypersurfaces given by quadratic equations of the form

$$(x^1 - a^1)^2 + \cdots + (x^n - a^n)^2 + \operatorname{sgn}(k)(x^{n+1} - a^{n+1})^2 = \text{constants},$$

where $a = (a^1, \ldots, a^{n+1})$ is an arbitrary fixed vector in E^{n+1}. In $R^n(0)$, these are just the usual hyperspheres. Among these hyperspheres the *great hyperspheres* are those sections of hyperplanes which pass through the center $(0, \ldots, 0, \operatorname{sgn}(k)/\sqrt{|k|})$ of $R^n(k)$ in E^{n+1}, $k \neq 0$. For $k = 0$, we consider the point at infinite on the x^{n+1}-axis as the center in E^{n+1}. The intersection of a hyperplane through the center in E^{n+1} is just a hyperplane in $R^n(0)$. All other hyperspheres in $R^n(k)$ are called *small hyperspheres*. Finally we shall mention that a hypersurface of $R^n(k)$ is a great hypersphere if and only if it is complete and totally geodesic (for the definition of totally geodesic submanifolds, see Chapter 2).

5 Conformal Changes of Riemannian Metric

Let M be an n-dimensional Riemannian manifold with metric tensor g and ρ a positive function on M. Then

$$g^* = \rho^2 g$$

defines a new metric tensor on M which does not change the angle between two vectors at a point. Hence, it is a *conformal change* of the metric. In particular, if the function ρ is a constant, the conformal transformation is said to be *homothetic*.

Let ∇^* denote the operator of covariant differentiation with respect to the Christoffel symbols $\left\{ {h \atop ji} \right\}^*$ formed with g^*. Then we have

$$(\nabla^* - \nabla)(X, Y) = \omega(X)Y + \omega(Y)X - g(X, Y)U \tag{5.1}$$

for any vector fields X, Y, where ω is a 1-form given by

$$\omega = d \log \rho$$

and U is a vector field given by

$$g(U, X) = \omega(X),$$

or in the local components

$$\left\{ {h \atop ji} \right\}^* = \left\{ {h \atop ji} \right\} + \delta_j^h \rho_i + \delta_i^h \rho_j - g_{ji} \rho^h, \tag{5.2}$$

where

$$\rho_i = \nabla_i \log \rho, \quad \rho^h = \rho_t g^{th}.$$

Let K^* denote the curvature tensor of the Riemannian metric g^* and put

$$s(X, Y) = (\nabla_X \omega)(Y) - \omega(X)\omega(Y) + \frac{1}{2}\omega(U)g(X, Y), \qquad (5.3)$$

$$g(SX, Y) = s(X, Y). \qquad (5.4)$$

Then we have

$$K^*(X, Y)Z = K(X, Y)Z - s(Y, Z)X + s(X, Z)Y$$
$$- g(Y, Z)SX + g(X, Z)SY, \qquad (5.5)$$

where X, Y, and Z are any vector fields, or in local components

$$K^{*h}_{kji} = K^h_{kji} - \delta^h_k \rho_{ji} + \delta^h_j \rho_{ki} - \rho^h_k g_{ji} + \rho^h_j g_{ki}, \qquad (5.6)$$

where

$$\rho_{ji} = \nabla_j \rho_i - \rho_j \rho_i + \frac{1}{2}\rho_t \rho^t g_{ji};$$
$$\rho^h_k = \rho_{kt} g^{th}. \qquad (5.7)$$

From this equation, we have

$$L^* = L + s, \qquad (5.8)$$

where

$$L(X, Y) = -\frac{1}{n-2}R(X, Y) + \frac{r}{2(n-1)(n-2)}g(X, Y) \qquad (5.9)$$

and L^* is the corresponding tensor of type (0,2) associated with g^*, or in local components

$$L^*_{ji} = L_{ji} + \rho_{ji}, \qquad (5.10)$$

where

$$L_{ji} = -\frac{1}{n-2}K_{ji} + \frac{r}{2(n-1)(n-2)}g_{ji} \qquad (5.11)$$

and L^*_{ji} has a similar expression.

Thus, eliminating s, we find

$$C^* = C, \tag{5.12}$$

$$C(X, Y)Z = K(X, Y)Z + L(Y, Z)X - L(X, Z)Y + g(Y, Z)NX$$
$$- g(X, Z)NY, \tag{5.13}$$

$$g(NX, Y) = L(X, Y) \tag{5.14}$$

for any vector fields X, Y, Z, and C^* has a similar expression, or in local components

$$C_{kji}^{*h} = C_{kji}^h, \tag{5.15}$$

where

$$C_{kji}^h = K_{kji}^h + \delta_k^h L_{ji} - \delta_j^h L_{ki} + L_k^h g_{ji} - L_j^h g_{ki}, \tag{5.16}$$

$$L_k^h = L_{kt} g^{th}, \tag{5.17}$$

and C_{kji}^{*h} has a similar expression.

From (5.12) we see that the tensor field C is invariant under any conformal change of the metric. We call this tensor field the *Weyl conformal curvature tensor*, or simply *conformal curvature tensor*. This tensor C vanishes identically for $n = 3$. (Weyl, 1918, 1939).

We also have

$$D^*(X, Y, Z) = D(X, Y, Z) + \omega(C(X, Y)Z), \tag{5.18}$$

where

$$D(X, Y, Z) = (\nabla_X L)(Y, Z) - (\nabla_Y L)(X, Z), \tag{5.19}$$

and D^* has a similar expression, or in local components

$$D_{kji}^* = D_{kji} + C_{kji}^h \rho_h, \tag{5.20}$$

where

$$D_{kji} = \nabla_k L_{ji} - \nabla_j L_{ki}. \tag{5.21}$$

If a Riemannian metric g is conformally related to a Riemannian metric g^* which is locally flat, then the Riemannian manifold with the metric g is said to be *conformally flat* or *conformally euclidean*.

The following is a well-known theorem of Weyl.

Theorem 5.1 (Weyl, 1918, 1939). *A necessary and sufficient condition for a Riemannian manifold to be conformally flat is that*

$$C = 0 \quad for\ n > 3$$

and

$$D = 0 \quad for\ n = 3.$$

It should be noted that if M is conformally flat of dimension $n > 3$, then $D = 0$ is a consequence of $C = 0$. The C is called the *Cotton tensor*.

6 Theorem of Frobenius

Given a mapping x of a manifold M into another manifold N, the *differential* at a point $P \in M$ of x is the linear mapping x_* of the tangent space $T_P(M)$ at P into the tangent space $T_{x(P)}(N)$ at $x(P)$ defined as follows.

For each tangent vector $X \in T_P(M)$, $x_*(X)$ is the tangent vector at $x(P)$ given by

$$(x_*(X))(f) = X(f \circ x), \tag{6.1}$$

for any differentiable function f on N. When it is necessary to specify the point P, we use the notation $(x_*)_P$. The transpose of $(x_*)_P$ is a linear mapping of the dual space $T^*_{x(P)}(N)$ into the dual space $T^*_P(M)$. For any q-form ω' on N, we define a q-form $x^*\omega'$ on M by

$$(x^*\omega')(X_1, \ldots, X_q) = \omega'(x_*X_1, \ldots, x_*X_q), \tag{6.2}$$

for any vectors $X_1, \ldots, X_q$ at P. The exterior differentiation d commutes with x^*, that is

$$d(x^*\omega') = x^*(d\omega') \tag{6.3}$$

for any q-form ω' on N.

A mapping x of M into N is said to be of *rank* q at a point $P \in M$, if the dimension of the image $x_*(T_P(M))$ of $T_P(M)$ under x_* is q. If the rank of x at the point P is equal to the dimension of M, then $(x_*)_P$ is said to be *injective*.

A mapping x of M into N is called an *immersion* if $(x_*)_P$ is injective for every point P of M. In this case we say that M is *immersed* in N by x or that M is an *immersed submanifold* of N, or simply a *submanifold* of N. When an immersion x is injective, it is called an *imbedding*. In this case we say that M is *imbedded* in N by x or that M is an *imbedded submanifold* of N. An open subset M of a manifold N can be considered as a submanifold in a natural manner, and we call it an *open piece* of N.

A q-dimensional *distribution* on a manifold M is a mapping $\mathcal{D}$ defined on M which assigns to each point P in M a q-dimensional linear subspace $\mathcal{D}_P$ of $T_P(M)$. A q-dimensional distribution $\mathcal{D}$ is called *differentiable* if there are q differentiable vector fields on a neighborhood of P which, for each point Q in this neighborhood, form a basis of $\mathcal{D}_Q$. The set of these q vector fields is called a *local basis* of the distribution $\mathcal{D}$. A vector field X belongs to the distribution $\mathcal{D}$, and we write $X \in \mathcal{D}$ if, for any point $P \in M$, $X_P \in \mathcal{D}_P$. A distribution $\mathcal{D}$ is said to be *involutive* if, for all differentiable vector fields X, Y belonging to $\mathcal{D}$, we have $[X, Y] \in \mathcal{D}$. By a distribution we shall always mean a differentiable distribution.

An imbedded submanifold M' of M is called an integral manifold of the distribution $\mathcal{D}$ if $x_*(T_P(M')) = \mathcal{D}_P$ for all $P \in M'$, where x is the imbedding of M' in M. If there exists no integral submanifold of $\mathcal{D}$ which contains M', then M' is called a *maximal integral submanifold* of $\mathcal{D}$. A distribution $\mathcal{D}$ is said to be *integrable* if, for every point $P \in M$, there is an integral manifold of $\mathcal{D}$ containing P.

The classical theorem of Frobenius can be stated as follows.

Theorem 6.1. *An involutive distribution $\mathcal{D}$ on M is integrable. Furthermore, through every $P \in M$ there passes a unique maximal integral manifold of $\mathcal{D}$ and every other integral manifold containing P is an open submanifold of this maximal one.*

Let $\mathcal{D}$ be a q-dimensional distribution on M. We put

$$\Omega = \{1\text{-forms } \omega \colon \omega(X_P) = 0 \text{ for } X \in \mathcal{D}, P \in M\} \tag{6.4}$$

and let $I(\Omega)$ denote the ideal generated by Ω in the ring of exterior polynomials on M.

Theorem 6.1 of Frobenius can also be written as

Theorem 6.2. *The distribution $\mathcal{D}$ is involutive if and only if $d\Omega$ is contained in the ideal $I(\Omega)$, where the differential system Ω is given by (6.4).*

In local expression, Theorem 6.2 can also be stated as follows:

Theorem 6.3. *Let $\omega^1, \ldots, \omega^r$ be 1-forms in M, linearly independent at a point P. Suppose that there exist 1-forms θ_β^α; $\alpha, \beta = 1, \ldots, r$, satisfying*

$$d\omega^\alpha = \theta_\beta^\alpha \wedge \omega^\beta. \tag{6.5}$$

Then there are functions f_α^β, g^β defined in M around P satisfying

$$\omega^\alpha = f_\beta^\alpha dg^\beta. \tag{6.6}$$

In the following, by a *foliation* we mean an involutive distribution on a manifold M.

Problems

1. Prove that Euler's differential equations of the integral

$$\int_a^b \sqrt{g_{ji} \frac{dx^j}{dt} \frac{dx^i}{dt}} \, dt$$

are

$$\frac{d^2 x^h}{d^2 s} + \left\{ \begin{matrix} h \\ ji \end{matrix} \right\} \frac{dx^j}{ds} \frac{dx^i}{ds} = 0,$$

where s is the arc length, that is a parameter satisfying

$$g_{ji} \frac{dx^j}{ds} \frac{dx^i}{ds} = 1.$$

2. Let ω be a p-form. Prove that

$$d^2\omega = \delta^2\omega = 0.$$

3. Prove that if M is orientable and closed, then a p-form ω is harmonic if and only if $\Delta\omega = 0$, where the operator Δ is given by (2.23).

4. Prove that the Christoffel symbols under a coordinate transformation satisfy (3.38).

5. Prove that (2.13), (2.14), and (2.15) hold.

6. Prove that the expression

$$\nabla_X\nabla_Y Z - \nabla_Y\nabla_X Z - \nabla_{[X,Y]}Z$$

defines a tensor field of type $(1,3)$.

7. Prove that a manifold M of dimension n is orientable if and only if M admits a continuous, nonvanishing, globally defined n-form.

8. If ω is a p-form and φ is a q-form, prove that

$$d(\omega \wedge \varphi) = (d\omega) \wedge \varphi + (-1)^P \omega \wedge d\varphi.$$

9. Let $\Psi = (\omega_i^j)$ be a $q \times q$-matrix of 1-forms defined in a neighborhood of the origin 0, say, in E^n satisfying $d\Psi = \Psi^2$. Prove the following:

  (i) There exists a unique $q \times q$ nonsingular matrix A of functions satisfying

$$\Psi = -A^{-1}dA$$

  with the initial value $A_0 = I$ (identity matrix) at 0.

  (ii) If Ψ is skew-symmetric, then A is orthogonal.
  [Hint: Use the theorem of Frobenius.]

10. Prove the theorem of Weyl: A necessary and sufficient condition for a Riemannian manifold to be conformally flat is that

$$C = 0, \quad \text{for } n > 3$$

and

$$D = 0, \quad \text{for } n = 3.$$

11. Use the theorem of Frobenius to prove that if $\mathcal{D}$ is a q-dimensional involutive distribution, then for every point P there is a coordinate neighborhood with the local coordinates $x^1, \ldots, x^n$ such that the coordinate slices $x^{q+1} = $ constant, $\ldots, x^n = $ constant are integral manifolds of $\mathcal{D}$.

12. Prove the second Bianchi identity:

$$(\nabla_X K)(Y, Z) + (\nabla_Y K)(Z, X) + (\nabla_Z K)(X, Y) = 0.$$

13. Let g be a nondegenerate symmetric tensor field of type (0,2) and of differentiability class C^∞, on a manifold M. Then g is called a *pseudo-Riemannian metric tensor* on M. A manifold with such a tensor is called a *pseudo-Riemannian manifold*. Prove that there exists a unique Riemannian connection on a pseudo-Riemannian manifold, that is, a metric connection without torsion.

14. Let M be a surface with *isothermal coordinates* x^1, x^2 such that

$$g = \lambda\{(dx^1)^2 + (dx^2)^2\}.$$

Prove that the Gaussian curvature G is given by

$$G = \frac{-1}{2\lambda}\Delta \log \lambda.$$

15. Prove that every complete flat surface is parabolic.

16. Prove that if the universal covering surface of a surface M is conformally equivalent to a euclidean plane, then the surface M is parabolic.

Chapter 2

Submanifolds

1 Induced Connection and Second Fundamental Form

Let M be an n-dimensional manifold immersed in an m-dimensional manifold N with $m > n$. Since the discussion is local, we may assume, if we want, that M is imbedded in N. If the manifold N is covered by a system of coordinate neighborhoods $\{V; u^A\}$ and M is covered by a system of coordinate neighborhoods $\{U; x^h\}$, where, here and in the sequel the indices $A, B, C, \ldots$ run over the range $1, 2, \ldots, m$ and $i, j, h, \ldots$ run over the range $1, 2, \ldots, n$, then the submanifold M can be represented locally by

$$u^A = u^A(x^h). \tag{1.1}$$

In the following, we shall identify vector fields in M and their images under the differential mapping, that is, if i denotes the immersion of M in N and X is a vector field in M, we identify X and $i_*(X)$. Thus, if X is a vector field in M and has the local expression $X = X^h \partial_h$ where $\partial_h = \partial/\partial x^h$, then X also has the local expression $X = B_h^A X^h \partial_A$ in N, where $\partial_A = \partial/\partial u^A$ and

$$B_h^A = \partial u^A/\partial x^h. \tag{1.2}$$

Let X be a vector field on M. Then a vector field $\tilde{X}$ defined on N is called an *extension* of X if its restriction to the submanifold M is X.

Suppose that the manifold N is a Riemannian manifold with Riemannian metric $\tilde{g}$, then the submanifold M is also a Riemannian manifold with the Riemannian metric g given by

$$g(X, Y) = \tilde{g}(X, Y)$$

for any vector fields X, Y in M. The Riemannian metric g on M is called the *induced metric* on M. In local components, $g_{ji} = g_{BA} B_j^B B_i^A$ with $g = g_{ji} dx^j dx^i$ and $\tilde{g} = g_{BA} du^B du^A$.

If a vector ξ_P of N at a point $P \in M$ satisfies

$$\tilde{g}(X_P, \xi_P) = 0, \tag{1.3}$$

for any vector X_P of M at P, then ξ_P is called a normal vector of M in N at P. A unit normal vector field of M in N is sometimes called a *normal section* on M, or a *normal direction* on M.

Let $T^\perp(M)$ denote the vector bundle of all normal vectors of M in N. Then the tangent bundle of N, restricted to M, is the direct sum of the tangent bundle $T(M)$ of M and the normal bundle $T^\perp(M)$ of M in N, that is,

$$T(N)|_M = T(M) \oplus T^\perp(M). \tag{1.4}$$

If the submanifold M is of codimension one in N and M and N are both orientable, we can always choose a normal section ξ on M, that is,

$$\tilde{g}(X, \xi) = 0, \quad \tilde{g}(\xi, \xi) = 1, \tag{1.5}$$

where X is any arbitrary vector field on M.

We denote by $\tilde{\nabla}$ the Riemannian connection on N with respect to its Riemannian metric $\tilde{g}$. The Riemannian connection $\tilde{\nabla}$ has no torsion, that is,

$$\tilde{\nabla}_{\tilde{X}} \tilde{Y} - \tilde{\nabla}_{\tilde{Y}} \tilde{X} - [\tilde{X}, \tilde{Y}] = 0, \tag{1.6}$$

and is *metric*, that is,

$$\tilde{\nabla}_{\tilde{X}}(\tilde{g}(\tilde{Y}, \tilde{Z})) = \tilde{g}(\tilde{\nabla}_{\tilde{X}} \tilde{Y}, \tilde{Z}) + \tilde{g}(\tilde{Y}, \nabla_{\tilde{X}} \tilde{Z}), \tag{1.7}$$

where $\tilde{X}$, $\tilde{Y}$, and $\tilde{Z}$ are arbitrary vector fields on N.

Proposition 1.1. *Let X and Y be two vector fields on M and let $\tilde{X}$ and $\tilde{Y}$ be extensions of X and Y, respectively. Then $[\tilde{X}, \tilde{Y}]|_M$ is independent of the extensions, and*

$$[\tilde{X}, \tilde{Y}]|_M = [X, Y]. \tag{1.8}$$

Proof. Let $\tilde{X}^A(u)$ and $\tilde{Y}^A(u)$ be extensions of $B_h^A X^h(u)$ and $B_h^A Y^h(u)$, respectively. Then we have

$$\tilde{X}^A(u(x)) = B_h^A X^h(u), \quad \tilde{Y}^A(u(x)) = B_h^A Y^h(u). \tag{1.9}$$

Consequently, $[\tilde{X}, \tilde{Y}]|_M$ has the components

$$
\begin{aligned}
[\tilde{X}, \tilde{Y}]^A|_{u=u(x)} &= (\tilde{X}^B \partial_B \tilde{Y}^A - \tilde{Y}^B \partial_B \tilde{X}^A)|_{u=u(x)} \\
&= (B_h^B X^h \partial_B \tilde{Y}^A - B_h^B Y^h \partial_B \tilde{X}^A)|_{u=u(x)} \\
&= X^h \partial_h (B_i^A Y^i) - Y^h \partial_h (B_i^A X^i) \\
&= B_i^A (X^h \partial_h Y^i - Y^h \partial_h X^i) \\
&= B_i^A [X, Y]^i.
\end{aligned}
$$

This shows that $[\tilde{X}, \tilde{Y}]|_M$ does not depend on the extensions $\tilde{X}$ and $\tilde{Y}$ of X and Y and is equal to $[X, Y]$. $\qquad\qquad\square$

Proposition 1.2. *Let X and Y be two vector fields on M and let $\tilde{X}$ and $\tilde{Y}$ be extensions of X and Y, respectively. Then $(\tilde{\nabla}_{\tilde{X}} \tilde{Y})|_M$ does not depend on the extensions. Denoting this by $\tilde{\nabla}_X Y$,*

$$
\tilde{\nabla}_X Y = \nabla_X Y + h(X, Y), \qquad (1.10)
$$

where ∇ is the Riemannian connection defined on the submanifold M with respect to g and $h(X, Y)$ is a normal vector field on M and is symmetric and bilinear in X and Y.

Proof. Let $\left\{{}_{BA}^{C}\right\}$ denote the components of the Riemannian connection $\tilde{\nabla}$. Then the components of $\tilde{\nabla}_{\tilde{X}} \tilde{Y}$ are given by

$$
\tilde{X}^B \left(\partial_B \tilde{Y}^C + \left\{ {}^{C}_{BA} \right\} \tilde{Y}^A \right),
$$

where $\tilde{X}^A$ and $\tilde{Y}^A$ are the local components of $\tilde{X}$ and $\tilde{Y}$, respectively. Hence we have

$$
\left. \tilde{X}^B \left(\partial_B \tilde{Y}^C + \left\{ {}^{C}_{BA} \right\} \tilde{Y}^A \right) \right|_{u=u(x)} = \left. B_i^B X^i \left(\partial_B \tilde{Y}^C + \left\{ {}^{C}_{BA} \right\} B_j^A Y^j \right) \right|_{u=u(x)}
$$

$$
= X^i \left(\partial_i (B_j^C Y^j) + \left\{ {}^{C}_{BA} \right\} B_i^B B_j^A Y^j \right).
$$

Therefore $\tilde{\nabla}_{\tilde{X}} \tilde{Y}$ does not depend on the extensions. From (1.4), we see that $\tilde{\nabla}_X Y$ can be expressed in the form

$$
\tilde{\nabla}_X Y = \nabla_X Y + h(X, Y),
$$

where $\nabla_X Y$ is a vector field tangent to M and $h(X, Y)$ is a vector field normal to M. Replacing X and Y by αX and βY, respectively, α and β being functions on M, we have

$$
\begin{aligned}
\tilde{\nabla}_{\alpha X}(\beta Y) &= \alpha \tilde{\nabla}_X (\beta Y) = \alpha \{ (X\beta)Y + \beta (\tilde{\nabla}_X Y) \} \\
&= \{ \alpha (X\beta) Y + \alpha\beta \nabla_X Y \} + \alpha\beta h(X, Y).
\end{aligned}
$$

Thus we find

$$\nabla_{\alpha X}(\beta Y) = \alpha(X\beta)Y + \alpha\beta\nabla_X Y$$

and

$$h(\alpha X, \beta Y) = \alpha\beta h(X, Y).$$

The first equation shows that ∇ defines an affine connection on M and the second equation shows that h is bilinear in X and Y, since additivity is trivial.

Since the Riemannian connection $\tilde{\nabla}$ has no torsion, we have, from Proposition 1.1, that

$$\begin{aligned}
0 &= \tilde{\nabla}_X Y - \tilde{\nabla}_Y X - [X, Y] \\
&= \nabla_X Y + h(X, Y) - \nabla_Y X - h(Y, X) - [X, Y],
\end{aligned}$$

from which, by comparing the tangential and normal parts, we find

$$\nabla_X Y - \nabla_Y X - [X, Y] = 0$$

and

$$h(X, Y) = h(Y, X).$$

These equations show that ∇ has no torsion and h is symmetric.

Since the Riemannian connection $\tilde{\nabla}$ is metric, we have

$$\begin{aligned}
\nabla_X(g(Y, Z)) = \tilde{\nabla}_X(\tilde{g}(Y, Z)) &= \tilde{g}(\tilde{\nabla}_X Y, Z) + \tilde{g}(Y, \tilde{\nabla}_X Z) \\
&= \tilde{g}(\nabla_X Y + h(X, Y), Z) + \tilde{g}(Y, \nabla_X Z + h(X, Z)) \\
&= \tilde{g}(\nabla_X Y, Z) + \tilde{g}(Y, \nabla_X Z) \\
&= g(\nabla_X Y, Z) + g(Y, \nabla_X Z)
\end{aligned}$$

for any vector fields X, Y, and Z on M. This shows that ∇ is the Riemannian connection of the induced metric g. This proves the proposition. $\square$

We call the Riemannian connection ∇ the *induced connection* and h the *second fundamental form* of the submanifold M.

Let ξ be a normal vector field on M and X be a vector field on M. We may then decompose $\tilde{\nabla}_X \xi$ as

$$\tilde{\nabla}_X \xi = -A_\xi(X) + \nabla_X^\perp \xi, \tag{1.11}$$

where $-A_\xi(X)$ and $\nabla_X^\perp \xi$ are, respectively, the tangential component and the normal component of $\tilde{\nabla}_X \xi$. It is easy to check that the vector fields $A_\xi(X)$ and $\nabla_X^\perp \xi$ are differentiable on M. It is also easy to verify that, for each point P in M, $h(X, Y)$ at P depends only on X_P and Y_P.

Proposition 1.3. (a) $A_\xi(X)$ is bilinear in ξ and X and hence $A_\xi(X)$ at a point $P \in M$ depends only on ξ_P and X_p.

(b) For each normal vector field ξ on M, we have

$$g(A_\xi(X), Y) = \tilde{g}(h(X, Y), \xi) \tag{1.12}$$

for any vector fields X and Y on M.

Proof. (a) For any function α and β on M, we have

$$\tilde{\nabla}_{\alpha X}(\beta\xi) = \alpha\tilde{\nabla}_X(\beta\xi) = \alpha\{(X\beta)\xi + \beta\tilde{\nabla}_X\xi\}$$
$$= \alpha(X\beta)\xi - \alpha\beta A_\xi(X) + \alpha\beta\nabla_X^\perp\xi.$$

This implies that

$$A_{\beta\xi}(\alpha X) = \alpha\beta A_\xi(X) \tag{1.13}$$

and

$$\nabla_{\alpha X}^\perp \beta\xi = \alpha(X\beta)\xi + \alpha\beta\nabla_X^\perp\xi. \tag{1.14}$$

Thus $A_\xi(X)$ is bilinear in ξ and X, since additivity is trivial.

(b) Let Y be an arbitrary vector field on M. Then we have

$$0 = \tilde{g}(\tilde{\nabla}_X Y, \xi) + \tilde{g}(Y, \tilde{\nabla}_X\xi)$$
$$= \tilde{g}(\nabla_X Y, \xi) + \tilde{g}(h(X, Y), \xi) - g(Y, A_\xi(X)) + \tilde{g}(Y, \nabla_X^\perp\xi)$$
$$= \tilde{g}(h(X, Y), \xi) - g(Y, A_\xi(X)),$$

since $\tilde{g}(\nabla_X Y, \xi) = \tilde{g}(Y, \nabla_X^\perp\xi) = 0$. This proves the proposition. $\square$

Proposition 1.4. $\nabla^\perp$ *is a metric connection in the normal bundle $T^\perp(M)$ of M in N with respect to the induced metric on $T^\perp(M)$.*

Proof. From (1.14) we see that $\nabla^\perp$ defines an affine connection on the normal bundle $T^\perp(M)$. Moreover, for any normal vector fields ξ and η in $T^\perp(M)$, we have

$$\tilde{\nabla}_X\xi = -A_\xi(X) + \nabla_X^\perp\xi, \quad \tilde{\nabla}_X\eta = -A_\eta(X) + \nabla_X^\perp\eta.$$

Hence, we have

$$\tilde{g}(\nabla_X^\perp\xi, \eta) + \tilde{g}(\xi, \nabla_X^\perp\eta) = \tilde{g}(\tilde{\nabla}_X\xi, \eta) + \tilde{g}(\xi, \tilde{\nabla}_X\eta)$$
$$= \tilde{\nabla}_X\tilde{g}(\xi, \eta) = \nabla_X^\perp\tilde{g}(\xi, \eta).$$

This shows that the connection $\nabla^\perp$ is metric for the fibre metric in the normal bundle induced from $\tilde{g}$. This proves the proposition. $\square$

A normal vector field ξ on M is said to be *parallel in the normal bundle*, or simply *parallel*, if we have $\nabla^\perp\xi = 0$ identically.

Formula (1.10) is called Gauss' formula, and formula (1.11) is called Weingarten's formula.

A submanifold M is said to be *totally geodesic* if the second fundamental form h vanishes identically, that is,

$$h = 0. \tag{1.15}$$

For a normal section ξ on M, if A_ξ is everywhere proportional to the identity transformation I, that is,

$$A = \rho I \tag{1.16}$$

for some function ρ, then ξ is called an *umbilical section* on M, or M is said to be umbilical with respect to ξ. If the submanifold M is umbilical with respect to every local normal section in M, then M is said to be *totally umbilical*.

Let $\xi_1, \ldots, \xi_{m-n}$ be an orthonormal basis of the normal space $T_P^\perp(M)$ at a point $P \in M$ and let $A^x = A_{\xi_x}$, then

$$H = \frac{1}{n}(\text{trace } A^x)\xi_x \tag{1.17}$$

is a normal vector at P which is independent of the choice of the orthonormal basis ξ_x, here and in the sequel, the indices x, y, z run over the range $1, \ldots, m - n$. We call the vector H the *mean curvature vector* at P.

A submanifold M is called a *minimal submanifold* if the mean curvature vector vanishes identically. If there exists a function λ on the submanifold M such that

$$\tilde{g}(h(X, Y), H) = \lambda g(X, Y) \tag{1.18}$$

for any vector fields X, Y on M, then the submanifold is called a *pseudoumbilical submanifold*. It is clear that every minimal submanifold is a pseudoumbilical submanifold, and for a pseudoumbilical submanifold with (1.18), $\lambda = \tilde{g}(H, H)$. A totally umbilical submanifold is totally geodesic if and only if it is minimal.

For a unit normal vector ξ of M at a point P, the second fundamental tensor with respect to ξ, A_ξ, is self-adjoint. Hence, there exist orthonormal vectors $E_1, \ldots, E_n$ of M at P which are the eigenvectors of A_ξ, that is,

$$A_\xi(E_i) = h_i E_i,$$

for real numbers h_i. We call the eigenvalues h_i the *principal curvatures* and the eigenvectors E_i the *principal directions* of the normal direction ξ.

2   Equations of Gauss, Codazzi, and Ricci

The curvature tensor of the Riemannian manifold N is given by

$$\tilde{K}(\tilde{X}, \tilde{Y})\tilde{Z} = \tilde{\nabla}_{\tilde{X}}\tilde{\nabla}_{\tilde{Y}}\tilde{Z} - \tilde{\nabla}_{\tilde{Y}}\tilde{\nabla}_{\tilde{X}}\tilde{Z} - \tilde{\nabla}_{[\tilde{X},\tilde{Y}]}\tilde{Z}, \tag{2.1}$$

for any vector fields $\tilde{X}$, $\tilde{Y}$, and $\tilde{Z}$ in N. Let X, Y, and Z be vector fields on the submanifold M. Then we have

$$\tilde{K}(X, Y)Z = \tilde{\nabla}_X \tilde{\nabla}_Y Z - \tilde{\nabla}_Y \tilde{\nabla}_X Z - \tilde{\nabla}_{[X,Y]}Z. \tag{2.2}$$

Thus, by Gauss' formula (1.10), we find

$$\begin{aligned}
\tilde{K}(X, Y)Z &= \tilde{\nabla}_X(\nabla_Y Z + h(Y, Z)) - \tilde{\nabla}_Y(\nabla_X Z + h(X, Z)) \\
&\quad - (\nabla_{[X,Y]}Z + h([X, Y], Z)) \\
&= \nabla_X \nabla_Y Z + h(X, \nabla_Y Z) + \tilde{\nabla}_X h(Y, Z) - \nabla_Y \nabla_X Z - h(Y, \nabla_X Z) \\
&\quad - \tilde{\nabla}_Y h(X, Z) - \nabla_{[X,Y]}Z - h([X, Y], Z) \\
&= K(X, Y)Z + h(X, \nabla_Y Z) - h(Y, \nabla_X Z) - h([X, Y], Z) \\
&\quad + \tilde{\nabla}_X h(Y, Z) - \tilde{\nabla}_Y h(X, Z),
\end{aligned}$$

where K denotes the curvature tensor of the submanifold M. Let $\xi_1, \dots, \xi_{m-n}$ be orthonormal normal vector fields of M and let h^x be the corresponding second fundamental forms, that is,

$$h(X, Y) = h^x(X, Y)\xi_x, \tag{2.3}$$

where h is the second fundamental form of the submanifold M. Then, from (1.11) and (1.12), we have

$$\begin{aligned}
\tilde{K}(X, Y)Z &= K(X, Y)Z + h^x(X, \nabla_Y Z)\xi_x - h^x(Y, \nabla_X Z)\xi_x - h^x([X, Y], Z)\xi_x \\
&\quad + X(h^x(Y, Z))\xi_x - h^x(Y, Z)A_x(X) + h^x(Y, Z)\nabla_X^\perp \xi_x \\
&\quad - Y(h^x(X, Z))\xi_x + h^x(X, Z)A_x(Y) - h^x(X, Z)\nabla_Y^\perp \xi_x \\
&= K(X, Y)Z - \{(\nabla_Y h^x)(X, Z) - (\nabla_X h^x)(Y, Z)\}\xi_x \\
&\quad - h^x(Y, Z)A_x(X) + h^x(X, Z)A_x(Y) + h^x(Y, Z)\nabla_X^\perp \xi_x \\
&\quad - h^x(X, Z)\nabla_Y^\perp \xi_x, \tag{2.4}
\end{aligned}$$

by virtue of

$$Xh^x(Y, Z) = (\nabla_x h^x)(Y, Z) + h^x(\nabla_X Y, Z) + h^x(Y, \nabla_X Z), \tag{2.5}$$

where $A_x = A^x$. Thus, for any vector field W on M, we have

$$\begin{aligned}
\tilde{K}(X, Y; Z, W) &= K(X, Y; Z, W) + \tilde{g}(h(X, Z), h(Y, W)) \\
&\quad - \tilde{g}(h(X, W), h(Y, Z)). \tag{2.6}
\end{aligned}$$

Moreover, from (2.4), we see that the normal component of $\tilde{K}(X, Y)Z$ is given by

$$\begin{aligned}
(\tilde{K}(X, Y)Z)^N &= \{(\nabla_X h^x)(Y, Z) - (\nabla_Y h^x)(X, Z)\}\xi_x + h^x(Y, Z)\nabla_X^\perp \xi_x \\
&\quad - h^x(X, Z)\nabla_Y^\perp \xi_x. \tag{2.7}
\end{aligned}$$

Equation (2.6) is called the *equation of Gauss* and Eq. (2.7) is called the *equation of Codazzi*.

For the second fundamental form h, if we define the covariant derivative, denoted by $\bar{\nabla}_X h$, to be

$$(\bar{\nabla}_X h)(Y, Z) = \nabla_X^\perp (h^*(Y, Z)\xi_x) - \{h^*(\nabla_X Y, Z) + h^*(Y, \nabla_X Z)\}\xi_x$$
$$= (\nabla_X h^*)(Y, Z)\xi_x + h^*(Y, Z)\nabla_X^\perp \xi_x, \tag{2.8}$$

then from this equation we see that the equation of Codazzi may also be written as

$$(\tilde{K}(X, Y)Z)^N = (\bar{\nabla}_X h)(Y, Z) - (\bar{\nabla}_Y h)(X, Z). \tag{2.9}$$

Moreover, if ξ and η are two normal vector fields of M, then, from (2.1), we have

$$\tilde{K}(X, Y; \xi, \eta) = \tilde{g}(\tilde{\nabla}_X \tilde{\nabla}_Y \xi, \eta) - \tilde{g}(\tilde{\nabla}_Y \tilde{\nabla}_X \xi, \eta) - \tilde{g}(\tilde{\nabla}_{[X,Y]}\xi, \eta)$$
$$= -\tilde{g}(\tilde{\nabla}_X(A_\xi(Y)), \eta) + \tilde{g}(\tilde{\nabla}_X \nabla_Y^\perp \xi, \eta) + \tilde{g}(\tilde{\nabla}_Y(A_\xi(X)), \eta)$$
$$- \tilde{g}(\tilde{\nabla}_Y \nabla_X^\perp \xi, \eta) - \tilde{g}(\nabla_{[X,Y]}^\perp \xi, \eta)$$
$$= -\tilde{g}(h(X, A_\xi(Y)), \eta) + \tilde{g}(h(Y, A_\xi(X)), \eta) + \tilde{g}(\tilde{\nabla}_X^\perp \nabla_Y^\perp \xi, \eta)$$
$$- \tilde{g}(\tilde{\nabla}_Y^\perp \nabla_X^\perp \xi, \eta) - \tilde{g}(\nabla_{[X,Y]}^\perp \xi, \eta).$$

Thus, if we denote by K^N the curvature tensor of the normal connection $\nabla^\perp$ on the normal bundle $T^\perp(M)$, that is,

$$K^N(X, Y)\xi = \nabla_X^\perp \nabla_Y^\perp \xi - \nabla_Y^\perp \nabla_X^\perp \xi - \nabla_{[X,Y]}^\perp \xi, \tag{2.10}$$

then we have

$$\tilde{K}(X, Y; \xi, \eta) = K^N(X, Y; \xi, \eta) + g(A_\eta A_\xi(X), Y) - g(A_\xi A_\eta(X), Y)$$
$$= K^N(X, Y; \xi, \eta) - g([A_\xi, A_\eta](X), Y), \tag{2.11}$$

where

$$K^N(X, Y; \xi, \eta) = \tilde{g}(K^N(X, Y)\xi, \eta) \tag{2.12}$$

and

$$[A_\xi, A_\eta] = A_\xi A_\eta - A_\eta A_\xi. \tag{2.13}$$

Equation (2.11) is called the *equation of Ricci*.

If the ambient space N is a space of constant curvature k, then we have

$$\tilde{K}(\tilde{X}, \tilde{Y})\tilde{Z} = k\{\tilde{g}(\tilde{Z}, \tilde{Y})\tilde{X} - \tilde{g}(\tilde{Z}, \tilde{X})\tilde{Y}\}, \tag{2.14}$$

for any vector fields $\tilde{X}, \tilde{Y}, \tilde{Z}$ in M. Hence, for any vector fields $X, Y,$ and Z in M, $\tilde{K}(X, Y)Z$ is tangent to M. Thus, the equation of Gauss and the equation of Ricci reduce to

$$K(X, Y; Z, W) = k\{g(X, W)g(Y, Z) - g(X, Z)g(Y, W)\}$$
$$+ \tilde{g}(h(X, W), h(Y, Z)) - \tilde{g}(h(X, Z), h(Y, W)) \tag{2.15}$$

and

$$K^N(X, Y; \xi, \eta) = g([A_\xi, A_\eta](X), Y), \tag{2.16}$$

or

$$\tilde{g}(\nabla_X^\perp \nabla_Y^\perp \xi - \nabla_Y^\perp \nabla_X^\perp \xi - \nabla_{[X,Y]}^\perp \xi, \eta) = g([A_\xi, A_\eta](X), Y), \tag{2.17}$$

respectively.

The equation of Codazzi reduces to

$$0 = (\overline{\nabla}_X h)(Y, Z) - (\overline{\nabla}_Y h)(X, Z). \tag{2.18}$$

In the following, we call a q-plane bundle over a manifold M a *Riemannian q-plane bundle* if it is equipped with a bundle metric and a compatible connection. If E is any q-plane bundle over a Riemannian manifold M, a *second fundamental tensor in E* is a section A in $\text{Hom}(T \otimes E, T)$ satisfying

$$g(A(X, \xi), Y) = g(X, A(Y, \xi)) \tag{2.19}$$

for any vector fields X, Y in M and ξ in E, where g is the metric tensor of M and $T = T(M)$ is the tangent bundle of M. If E is a Riemannian vector bundle with a second fundamental tensor A, we define the associated second fundamental form h by

$$\tilde{g}(h(X, Y), \xi) = g(A(X, \xi), Y) \tag{2.20}$$

for any vector fields X, Y in M and ξ in E, where $\tilde{g}$ is the bundle metric tensor of E.

We can now state the *fundamental theorems of submanifolds* as follows. For the proofs, see, for instance, Bishop and Crittenden (1964).

Existence Theorem. *Let M be a simply connected n-dimensional Riemannian manifold with a Riemannian q-plane bundle E over M equipped with a second fundamental form h and associated second fundamental tensor A. If they satisfy the equation (2.15) of Gauss, the equation (2.17) of Ricci, and the equation (2.18) of Codazzi, then M can be isometrically immersed in an $(n + q)$-dimensional space form $R^{n+q}(k)$ of curvature k with normal bundle E.*

Rigidity Theorem. *Let x, x': $M \to R^m(k)$ be two isometric immersions of an n-dimensional Riemannian manifold M in an m-dimensional space form $R^m(k)$ of curvature k with normal bundles E, E' equipped with their canonical bundle metrics, connections, and second fundamental forms. Suppose that there is an isometry $f: M \to M$ such that f can be covered by a bundle map $\overline{f}: E \to E'$ which preserves the bundle metrics, the connections, and the second fundamental forms. Then there is a rigid motion F of $R^m(k)$ such that $F \circ x = x' \circ f$.*

3   Totally Umbilical Submanifolds

In this section, we shall consider totally umbilical submanifolds in a space of constant curvature.

Proposition 3.1. *A totally umbilical submanifold M in a space N of constant curvature k is also of constant curvature.*

Proof. Let ξ_x be an orthonormal basis in the normal bundle. Then we have

$$g(A_x(X), Y) = \tilde{g}(h(X, Y), \xi_x) = \lambda_x g(X, Y)$$

for functions λ_x in M. This implies that

$$\text{trace } A_x = n\lambda_x, \qquad (3.1)$$

and

$$h(X, Y) = g(X, Y)\lambda^x \xi_x, \qquad (3.2)$$

where $\lambda^x = \lambda_x$. Therefore, we have

$$h(X, Y) = g(X, Y)H, \qquad (3.3)$$

where H is the mean curvature vector. Substituting this into (2.15), we find

$$K(X, Y; Z, W) = (k + |H|^2)\{g(X, W)g(Y, Z) - g(X, Z)g(Y, W)\}.$$

This shows that the submanifold M is of constant curvature $k + |H|^2$ for $n > 2$. If $n = 2$, $|H| = $ constant follows from the equation of Codazzi. This proves the proposition.   □

Proposition 3.2. *A totally umbilical submanifold M in a space form $R^m(k)$ is either totally geodesic in $R^m(k)$ or contained in a hypersphere of an $(n + 1)$-dimensional totally geodesic subspace of $R^m(k)$.*

Proof. From the proof of Proposition 3.1, we see that if M is totally umbilical submanifold in a space form $R^m(k)$ of curvature k, then the mean curvature vector H has constant length. If the mean curvature vector H vanishes, then M is totally geodesic. Hence, we may assume that the mean curvature H is nonzero. In this case, we may choose $m - n$ orthonormal normal vector fields ξ_x in M such that

$$H = |H|\xi_1. \qquad (3.4)$$

Since M is totally umbilical, the second fundamental forms with respect to $\xi_2, \ldots, \xi_{m-n}$ vanish, that is,

$$A_2 = \cdots = A_{m-n} = 0, \qquad (3.5)$$

by virtue of (3.4). Thus, from the equation of Codazzi and Eq. (3.3), we have

$$g(X, Z)\nabla_Y^\perp \xi_1 = g(Y, Z)\nabla_X^\perp \xi_1, \qquad (3.6)$$

from which, we find

$$\nabla_X^\perp \xi_1 = 0 \qquad (3.7)$$

for any vector field X in M. This shows that ξ_1 and H are parallel in the normal bundle. Thus, from (3.5), we see that the normal subspace spanned by $\xi_2, \ldots, \xi_{m-n}$

is invariant under parallel translation with respect to connection $\tilde{\nabla}$ of the space form $R^m(k)$, that is,

$$\tilde{\nabla}_X(\xi_2 \wedge \cdots \wedge \xi_{m-n}) = 0 \tag{3.8}$$

for any vector field X in M. We now consider the cases $k = 0$, $k > 0$, and $k < 0$ separately.

Case (i) $k = 0$. In this case the space $R^m(0) = E^m$ has a global parallelism, and all the tangent spaces can be identified with E^m itself. Equation (3.8) implies that the normal subspace spanned by $\xi_2, \ldots, \xi_{m-n}$ is independent of the choice of the point $P \in M$. Hence, the linear subspaces of E^m which are spanned by the tangent spaces and the mean curvature vector H are a fixed $(n + 1)$-dimensional linear subspace of E^m, say E^{n+1}.

Let

$$\tilde{X} = (X^1, \ldots, X^m)$$

be the position vector of E^m with the standard coordinates $X^1, \ldots, X^m$ in E^m. Then, by (3.3) and (3.7), we find

$$\tilde{\nabla}_Y\left(\tilde{X} + \frac{\xi_1}{|H|}\right) = \tilde{\nabla}_Y\tilde{X} - \frac{1}{|H|}A_1(Y) + \nabla_Y^{\perp}\left(\frac{\xi_1}{|H|}\right)$$
$$= Y - Y = 0.$$

for any vector field Y in M. This shows that the vector field

$$\tilde{X} + \frac{\xi_1}{|H|} = C,$$

restricted to M, is a constant vector in E^m. Thus, M is contained in a hypersphere S^{m-1} of E^m with radius $1/|H|$ and centered at $-C$. Since M is also contained in an $(n + 1)$-dimensional linear subspace of E^m, M must be contained in a hypersphere of an $(n + 1)$-dimensional totally geodesic subspace of E^m.

Case (ii) $k = 1$ [resp. Case (iii) $k = -1$]. For simplicity, we shall consider $R^m(1)$ [resp. $R^m(-1)$] as the elliptic space form (resp. hyperbolic space form) given in §4 of Chapter 1. We take the position vector $\tilde{X}$ relative to the center

$$(0, \ldots, 0, 1) \text{ [resp. } (0, \ldots, 0, -1)]$$

of $R^m(1)$ [resp. $R^m(-1)$] in E^{m+1}. For each point $P \in R^m(1)$ [resp. $R^m(-1)$], $\eta = \tilde{X}$ is a unit normal vector to $R^m(1)$ [resp. $R^m(-1)$] in E^{m+1} with respect to the Riemannian metric $\tilde{g}$ (resp. pseudo-Riemannian metric $\tilde{g}$) given by (I.4.2), where (I.4.2) means Eq. 4.2 of Chapter 1. It is easy to verify that $\nabla_W^*\eta = W$ for any vector field W in $R^m(1)$ [resp. $R^m(-1)$], where ∇^* is the euclidean connection on E^{m+1}

which is the Riemannian connection induced from the metric $g^* = (dx^1)^2 + \cdots + (dx^{m+1})^2$ [resp. $(dx^1)^2 + \cdots + (dx^m)^2 - (dx^{m+1})^2$]. Moreover, we have

$$\nabla_U^* V = \tilde{\nabla}_U V - g^*(U, V)\eta$$

for any vector fields U, V on $R^m(1)$ [resp. $R^m(-1)$]. In particular, we have

$$\nabla_X^* \xi_x = \tilde{\nabla}_X \xi_x, \quad x = 1, \ldots, m - n,$$

for any vector field X in M. Thus, the submanifold M is totally umbilical in E^{m+1}. In particular, we may conclude that M is contained in the intersection of an $(n + 1)$-dimensional linear subspace of E^{m+1} and $R^m(1)$ [resp. $R^m(-1)$]. From this, it is easy to see that M is contained in a hypersphere of an $(n + 1)$-dimensional totally geodesic subspace of $R^m(1)$ [resp. $R^m(-1)$]. $\qquad\square$

Remark 3.1. In the following, by an *n-sphere of a space form* $R^m(k)$ of curvature k, we mean a hypersphere of an $(n + 1)$-dimensional totally geodesic submanifold of $R^m(k)$. If the *n*-sphere of $R^m(k)$ is a great hypersphere (resp. small hypersphere) of an $(n + 1)$-dimensional totally geodesic submanifold of $R^m(k)$, then it is called a *great n-sphere* of $R^m(k)$ [resp. *small n-sphere* of $R^m(k)$]. From Proposition 3.2, we see that every totally umbilical submanifold M of a space form $R^m(k)$ is contained in an *n*-sphere of $R^m(k)$, where n is the dimension of the submanifold M.

In the following, by an *n*-dimensional *linear subspace of a euclidean m-space* E^m we mean a great *n*-sphere of E^m.

4 Scalar Curvature of Submanifolds

Let M be an n-dimensional manifold immersed in an m-dimensional Riemannian manifold N of constant curvature k. Let P be a point in M and x^i the local coordinates around P in M such that $X_i = \delta_i$ form an orthonormal basis of $T_P(M)$ at P. Let ξ_x be orthonormal normal vector fields of M. We put

$$h(X_i, X_j) = h^x(X_i, X_j)\xi_x = h_{ij}^x \xi_x. \qquad (4.1)$$

Then we have $h_{ji}^x = h_{ij}^x$. Let $\langle h \rangle$ denote the length of the second fundamental form h, that is,

$$\langle h \rangle^2 = h_{ji}^x h_x^{ji}, \qquad (4.2)$$

where $h_x^{ji} = g^{jt} g^{is} h_{ts}^x$.

From the equation of Gauss, we find that the scalar curvature r and the mean curvature vector H satisfy the following relation:

$$r = n^2 |H|^2 - \langle h \rangle^2 + n(n - 1)k. \qquad (4.3)$$

Theorem 4.1 (Chen and Okumura, 1973). *Let M be an n-dimensional sub-manifold of a space N of constant curvature k. If the scalar curvature r satisfies*

$$r \geq (n-2)\langle h \rangle^2 + (n-1)(n-2)k + 2(n-1)c \quad (resp. >) \tag{4.4}$$

at a point $P \in M$ for some number c, then the sectional curvatures of M are $\geq c$ (resp. > c) at the point P.

Proof. First we state the following lemma:

Lemma 4.1. *Let $a_1, \ldots, a_n$, b be $n+1$ $(n > 1)$ real numbers satisfying the following inequality.*

$$\left(\sum_{i=1}^{n} a_i \right)^2 \geq (n-1) \sum_{i=1}^{n} (a_i)^2 + b \quad (resp. >). \tag{4.5}$$

Then, for any distinct i and j, we have

$$2a_i a_j \geq \frac{b}{n-1} \quad (resp. >). \tag{4.6}$$

Proof of the Lemma. From (4.5) we find

$$(n-2)(a_n)^2 - 2 \left(\sum_{i=1}^{n-1} a_i \right) a_n$$

$$+ \left\{ (n-2) \sum_{i=1}^{n-1} (a_i)^2 - 2 \sum_{i<j \leq n-1} a_i a_j + b \right\} \leq 0, \quad (resp. <). \tag{4.7}$$

Denote the left-hand side of (4.7) by $-a$ $(a \geq 0)$. Then, since a_n is real, the discriminate of the quadratic equation

$$(n-2)(a_n)^2 - 2 \left(\sum_{i=1}^{n-1} a_i \right) a_n$$

$$+ \left\{ (n-2) \sum_{i=1}^{n-1} (a_i)^2 - 2 \sum_{i<j \leq n-1} a_i a_j + b + a \right\} = 0$$

is ≥ 0, that is,

$$\left(\sum_{i=1}^{n-1} a_i \right)^2 \geq (n-2) \left\{ (n-2) \sum_{i=1}^{n-1} (a_i)^2 - 2 \sum_{i<j \leq n-1} a_i a_j + b + a \right\}$$

$$\geq (n-2) \left\{ (n-1) \sum_{i=1}^{n-1} (a_i)^2 - \left(\sum_{i=1}^{n-1} a_i \right)^2 + b \right\} \quad (resp. >),$$

we have

$$\left(\sum_{i=1}^{n-1} a_i\right)^2 \geq (n-2)\sum_{i=1}^{n-1}(a_i)^2 + \left(\frac{n-2}{n-1}\right)b \quad \text{(resp. >).}$$

This inequality is of the same type as (4.5). Continuing the same process $(n-2)$ times, we obtain $2a_1a_2 \geq b/(n-1)$ (resp. >). This proves the lemma. $\qquad\square$

Now we return to the proof of the theorem. Substituting (4.3) into (4.4), we obtain

$$n^2|H|^2 \geq (n-1)\langle h\rangle^2 - 2(n-1)k + 2(n-1)c \quad \text{(resp. >).} \tag{4.8}$$

For simplicity, we may choose ξ_1 to be in the direction of the mean curvature vector H at P and $X_1, \ldots, X_n$ to be an orthonormal basis at P. (If $H = 0$ at P, we may choose an arbitrary ξ_1.) Then we have

$$h_{ji}^x = h_{jx}^i = h_x^{ji}$$

and

$$n^2|H|^2 = \left(\sum_{i=1}^n h_{ii}\right)^2 \tag{4.9}$$

at the point P, where $h_{ij} = h_{ij}^1$. From (4.8), we have

$$\left(\sum_{i=1}^n h_{ii}\right)^2 \geq (n-1)\sum_{i=1}^n (h_{ii})^2 + (n-1)\sum_{i\neq j}(h_{ij})^2$$

$$+ (n-1)\sum_{x=2}^{m-n} h_{ji}^x h_x^{ji} - 2(n-1)k + 2(n-1)c \quad \text{(resp. >).}$$

$$\tag{4.10}$$

Applying Lemma 4.1 to (4.10), we find

$$2h_{ii}h_{jj} - 2(h_{ij})^2 \geq \sum_{x=2}^{m-n} h_{kl}^x h_x^{kl} - 2k + 2c \quad \text{(resp. >).}$$

$$\geq \sum_{x=2}^{m-n}\{(h_{ii}^x)^2 + (h_{jj}^x)^2 + 2(h_{ij}^x)^2\} - 2k + 2c$$

$$\geq 2\sum_{x=2}^{m-n}\{|h_{ii}^x h_{jj}^x| + (h_{ij}^x)^2\} - 2k + 2c, \tag{4.11}$$

at the point P, for any distinct i and j. The sectional curvature of M for the plane section $\gamma(X_i, X_j)$ spanned by X_i and X_j is given by

$$k(\gamma(X_i, X_j)) = \{h_{ii}^x h_{jjx} - h_{ij}^x h_{ijx}\} + k, \tag{4.12}$$

at the point P, where $h_{ijx} = h_{ij}^x$. Thus, from (4.11), we find that the sectional curvatures of M at the point P are $\geqq c$ (resp. $> c$). This proves the theorem.   □

As a corollary of this theorem, we have immediately the following

Corollary 4.1 (Chen and Okumura, 1973). *Let M be an n-dimensional submanifold of a space N of constant curvature k. If the scalar curvature r satisfies*

$$r \geqq (n-2)\langle h \rangle^2 + (n-2)(n-1)k \quad (resp. \ >). \qquad (4.13)$$

at a point $P \in M$, then the sectional curvature of M is nonnegative (resp. positive) at the point P.

Since the scalar curvature is the sum of the sectional curvatures with respect to an orthonormal basis of the tangent space at a point, the scalar curvature is weaker than the sectional curvature in the sense of implication. From Theorem 4.1, we see that, under a certain suitable condition, we can also get some information about the sectional curvatures in terms of the scalar curvature for submanifolds in space forms. For Kaehlerian case, see Chen and Ogiue (1973).

5   Submanifolds of Euclidean Space or Sphere

In this section, we assume that the ambient space N is a euclidean m-space E^m with the position vector field $\tilde{X}$ with respect to the origin of E^m. Let M be an n-dimensional submanifold of E^m and ξ a unit normal vector field of M in E^m. The function $s(\xi)$, given by

$$s(\xi) = \tilde{g}(\tilde{X}, \xi),$$

is called the *support function* of M with respect to the normal vector field ξ.

For a unit normal vector field ξ of M in E^m, if the determinant of the second fundamental tensor A_ξ of ξ is nowhere zero, then ξ is called a *nondegenerate section* on M.

Proposition 5.1. *Let M be an n-dimensional submanifold of a euclidean m-space E^m. Then M is contained in a small hypersphere of E^m centered at the origin if and only if there exists a parallel nondegenerate section ξ on M such that the support function $s(\xi)$ with respect to ξ is constant.*

Proof. Let $E_1, \ldots, E_n$ be the principal directions of the unit normal vector field ξ with the principal curvatures $h_1, \ldots, h_n$. Suppose that the normal section ξ is nondegenerate and parallel in the normal bundle such that the support function $s(\xi)$ is a constant. Then we have

$$A_\xi(E_i) = h_i E_i.$$

Hence, we find

$$\tilde{\nabla}_{E_i} \xi = -h_i E_i$$

by the parallelism of ξ in the normal bundle. Thus,

$$0 = E_i \tilde{g}(\tilde{X}, \xi) = \tilde{g}(\tilde{X}, \tilde{\nabla}_{E_i}\xi) = -h_i \tilde{g}(\tilde{X}, E_i),$$

since $\tilde{\nabla}_{E_i}\tilde{X}$ has no normal component. Therefore, by the nondegeneracy of ξ, we find

$$\tilde{g}(\tilde{X}, E_1) = \cdots = \tilde{g}(\tilde{X}, E_n) = 0,$$

that is, the position vector $\tilde{X}$, restricted to M, is normal to M. Hence, we find

$$\tilde{\nabla}_{E_i}\tilde{g}(\tilde{X}, \tilde{X}) = 2\tilde{g}(\tilde{\nabla}_{E_i}\tilde{X}, \tilde{X}) = 0.$$

This shows that the submanifold M is contained in a small hypersphere of E^m centered at the origin. The converse of this is trivial. This proves the proposition. $\square$

For a submanifold M, the position vector field $\tilde{X}$ can be decomposed into two components:

$$\tilde{X} = \tilde{X}_t + \tilde{X}_n, \tag{5.1}$$

where $\tilde{X}_t$ is tangent to M and $\tilde{X}_n$ is normal to M. If $\tilde{X}_n$ is nowhere zero, we denote by η the unit normal vector field of M in the direction of $\tilde{X}_n$, that is,

$$\tilde{X}_n = f\eta, \quad f > 0. \tag{5.2}$$

We call the function f the *canonical support function* of M.

Proposition 5.2. *Let M be an n-dimensional submanifold of a euclidean m-space E^m. Then M is contained in a small hypersphere of E^m centered at the origin if and only if the canonical support function f is a nonzero constant and the determinant of A_η is not identically zero, where η is the normal section given by (5.2).*

Proof. If the submanifold M has nonzero constant canonical support function f and the determinant of A_η is not identically zero, then the normal direction η is well-defined. Let $E_1, \ldots, E_n$ be the principal directions of the normal direction η with the principal curvature $h_1, \ldots, h_n$. Then we have

$$\tilde{\nabla}_{E_i}\eta = -h_i E_i + \nabla^\perp_{E_i}\eta. \tag{5.3}$$

Since for any normal vector field ξ in M orthogonal to η we have

$$\tilde{g}(\tilde{X}, \xi) = \tilde{g}(\tilde{X}_n, \xi) = f\tilde{g}(\eta, \xi) = 0, \tag{5.4}$$

Eq. (5.3) implies that

$$0 = \tilde{\nabla}_{E_i} f = \tilde{\nabla}_{E_i}\tilde{g}(\tilde{X}, \eta) = \tilde{g}(\tilde{X}, \tilde{\nabla}_{E_i}\eta)$$
$$= -h_i \tilde{g}(\tilde{X}, E_i). \tag{5.5}$$

Let Ψ be the open subset of M such that

$$\Psi = \{P \in M : \det A_\eta = h_1 h_2 \cdots h_n \neq 0 \text{ at } P\}.$$

Then, Ψ is nonempty. From (5.5), we see that the position vector field X, restricted to M, is normal to Ψ. Hence, every component of Ψ is contained in a hypersphere of E^m centered at the origin. From this we see that the function

$$\det A_\eta = h_1 h_2 \cdots h_n$$

is a constant function on every component of Ψ. This implies that the open subset Ψ is also closed. Consequently, we have $\Psi = M$ and M is contained in a hypersphere of E^m centered at the origin. The converse of this is trivial. This proves the proposition.   □

Let M be an n-dimensional submanifold of a euclidean m-space E^m. It is clear that M cannot be contained in any proper subspace of the linear subspace of E^m spanned by the mean curvature vector H. On the other hand, if the submanifold M is closed, then we have the following proposition (see, for instance, Chen, 1971k; Kobayashi and Nomizu, 1969).

Proposition 5.3. *Let M be an n-dimensional closed submanifold of a euclidean m-space E^m. Then, for any fixed vector $C \neq 0$ in E^m, if we have $g(H, C) \geq 0$ or $g(H, C) \leq 0$ everywhere on M, M is contained in a hyperplane of E^m with C as its hyperplane normal vector in E^m, where a hyperplane of E^m means a great hypersphere of E^m.*

Proof. Let $\tilde{X}$ be the position vector field of E^m with respect to the origin. By a direct simple computation, we have

$$\Delta \tilde{X} = nH \tag{5.6}$$

on the submanifold M, where Δ is the Laplacian of M. Hence, for any fixed vector $C \neq 0$ in E^m, we have

$$\Delta \tilde{g}(X, C) = n\tilde{g}(H, C). \tag{5.7}$$

From (5.7) and Hopf's lemma, we see that, if we have either $\tilde{g}(\tilde{H}, C) \geq 0$ or $\tilde{g}(\tilde{H}, C) \leq 0$ everywhere, we have

$$\tilde{g}(\tilde{X}, C) = \text{constant}.$$

This implies that the submanifold M is contained in a hyperplane of E^m with C as its hyperplane normal vector in E^m. From this we obtain the proposition.   □

From Proposition 5.3, we have immediately the following

Corollary 5.1. *Let M be a closed submanifold of a euclidean m-space E^m. Then M is contained in the linear subspace of E^m spanned by the mean curvature vector.*

Corollary 5.2 (Chern and Hsiung, 1962/63). *There exists no closed minimal submanifold in a euclidean m-space.*

By using formula (5.6), we also have the following:

Proposition 5.4 (Takahashi, 1966). *Let M be an n-dimensional submanifold of a euclidean m-space E^m. Then M is a minimal submanifold of E^m if and only if $\Delta \tilde{X} = 0$, that is, all of the coordinate functions of M are harmonic.*

Remark 5.1. If the codimension is one, then Corollary 5.2 was first proved by Myers (1951).

Remark 5.2. Using the same method of Myers (1951), we may generalize Corollary 5.2 to the following:

Proposition 5.5. *There exists no closed minimal submanifold in a complete simply connected Riemannian manifold of nonpositive sectional curvature; in particular, there exists no closed minimal submanifold in a space form of nonpositive curvature.*

6 Connection of van der Waerden-Bortolotti

In §2 of this chapter, we defined the covariant derivative $\overline{\nabla} h$ of the second fundamental form h, which is given by

$$(\overline{\nabla}_X h)(Y, Z) = \nabla_X^\perp (h(Y, Z)) - h(\nabla_X Y, Z) - h(Y, \nabla_X Z). \qquad (6.1)$$

The equation of Codazzi is then given by

$$(\tilde{K}(X, Y)Z)^N = (\overline{\nabla}_X h)(Y, Z) - (\overline{\nabla}_Y h)(X, Z), \qquad (6.2)$$

where $(\tilde{K}(X, Y)Z)^N$ is the normal component of $\tilde{K}(X, Y)Z$. We may define the second covariant derivative $\overline{\nabla}_W \overline{\nabla}_X h$ of h by

$$(\overline{\nabla}_W \overline{\nabla}_X h)(Y, Z) = \nabla_W^\perp (\overline{\nabla}_X h(Y, Z)) - (\overline{\nabla}_X h)(\nabla_W Y, Z)$$
$$- (\overline{\nabla}_X h)(Y, \nabla_W Z) - (\overline{\nabla}_{\nabla_W X} h)(Y, Z) \qquad (6.3)$$

for any vector fields X, Y, Z, W in M. By direct computation, we may prove that $(\overline{\nabla}_W \overline{\nabla}_X h)(Y, Z)$ is a normal vector field of M, and we have

$$(\overline{\nabla}_W \overline{\nabla}_X h)(Y, Z) - (\overline{\nabla}_X \overline{\nabla}_W h)(Y, Z) = h(K(X, W)Y, Z) + h(Y, K(X, W)Z)$$
$$- K^N(X, W)(h(Y, Z)), \qquad (6.4)$$

where K^N is the curvature tensor of the normal connection $\nabla^\perp$.

The covariant derivative $\overline{\nabla}_X$ defined above is, in fact, the covariant derivative of the second fundamental form h with respect to the connections in the tangent bundle $T(M)$ and the normal bundle $T^\perp(M)$. This covariant differentiation can be extended to tensor fields of mixed type on M. For example, for a tensor field, say T, of type (0,3) with values in the normal bundle $T^\perp(M)$, we define

$$\overline{\nabla}_W T(X, Y, Z) = \nabla_W^\perp T(X, Y, Z) - T(\nabla_W X, Y, Z)$$
$$- T(X, \nabla_W Y, Z) - T(X, Y, \nabla_W Z). \tag{6.5}$$

Then the new tensor field $\overline{\nabla}T$ also has values in the normal bundle $T^\perp(M)$. The second fundamental form h is a tensor field of type (0,2) with values in $T^\perp(M)$ and $\overline{\nabla}h$ is the covariant derivative of h with values in $T^\perp(M)$. Similarly, $\overline{\nabla}\,\overline{\nabla}h$ is the covariant derivative of $\overline{\nabla}h$. We call the covariant derivative $\overline{\nabla}T$ of a tensor field T the *covariant derivative of van der Waerden-Bortolotti* of T, and the connection $\overline{\nabla}$ on the $T^\perp(M)$-valued tensor bundle of M the *connection of van der Waerden-Bortolotti*.

For convenience, we shall give the local expressions of the equations of Gauss, Codazzi, and Ricci as follows:

$$\tilde{K}_{kjih} = K_{kjih} - (h_{kh}^x h_{jix} - h_{jh}^x h_{kix}), \tag{6.6}$$

$$\tilde{K}_{kjih}^x = \overline{\nabla}_k h_{ji}^x - \overline{\nabla}_j h_{ki}^x, \tag{6.7}$$

$$\nabla_k l_{jy}^x - \nabla_j l_{ky}^x - h_{kt}^x h_{jy}^t + h_{jt}^x h_{ky}^t + l_{kz}^x l_{jy}^z - l_{jz}^x l_{ky}^z = \tilde{K}_{kjy}^x, \tag{6.8}$$

where $\nabla_j^\perp \xi_y = l_{jy}^x \xi_x$ with $l_{jy}^x = -l_{jx}^y$.

Moreover, Eq. (6.4) can be written as the local expression

$$\overline{\nabla}_l \overline{\nabla}_k h_{ji}^x - \overline{\nabla}_k \overline{\nabla}_l h_{ji}^x = K_{klj}^t h_{ti}^x + K_{kli}^t h_{tj}^x - K_{kly}^x h_{ji}^y. \tag{6.9}$$

In particular, if the ambient space N is of constant curvature k, then Eqs. (6.6), (6.7), and (6.8) reduce to

$$K_{kjih} = k(g_{kh}g_{ji} - g_{jh}g_{ki}) + (h_{kh}^x h_{jix} - h_{jh}^x h_{kix}), \tag{6.10}$$

$$0 = \overline{\nabla}_k h_{ji}^x - \overline{\nabla}_j h_{ki}^x, \tag{6.11}$$

$$\nabla_k l_{jy}^x - \nabla_j l_{ky}^x - h_{kt}^x h_{jy}^t + h_{jt}^x h_{ky}^t + l_{kz}^x l_{jy}^z - l_{jz}^x l_{ky}^z = 0. \tag{6.12}$$

Transvecting g^{kh} to (6.10), we find

$$K_{ji} = (n - 1)k g_{ji} + n h^{(x)} h_{jix} - h_{jt}^x h_{ix}^t, \tag{6.13}$$

where $h^{(x)} = (1/n)h_{tx}^t$ and K_{ji} is the Ricci tensor.

Moreover, from (2.16), we have

$$K_{jiy}^x = h_{jt}^x h_{iy}^t - h_{it}^x h_{jy}^t. \tag{6.14}$$

Problems

1. Let M be a totally umbilical submanifold of a space form. Use the equation of Codazzi to prove that the mean curvature vector has constant length.

2. Let M be an n-dimensional submanifold of a Riemannian manifold N and η a unit normal vector of M at a point $P \in M$. Let $h_1, \ldots, h_n$ be the principal curvatures of M with respect to η. Put

$$\binom{n}{l} M_l(\eta) = \sum h_1 \cdots h_l, \quad M_0(\eta) = 1,$$

where $\binom{n}{l} = n!/l!(n-l)!$. We call $M_l(\eta)$ the *l-th mean curvature with respect to* η. Prove that

(i) $M_l^2 - M_{l-1}M_{l+1} \geqq 0$ and equality holds if and only if $h_1 = \cdots = h_n$, where M_l denotes $M_l(\eta)$.

(ii) If $M_1, \ldots, M_l > 0$, then

$$M_1 \geqq (M_2)^{1/2} \geqq \cdots \geqq (M_l)^{1/l},$$

and the equality at any stage implies $h_1 = \cdots = h_n$.

(iii) If $M_1, \ldots, M_l > 0$, then

$$M_1 M_l - M_{l-1} \geqq 0,$$

and equality holds if and only if $h_1 = \cdots = h_n$.

3. Let M be an n-dimensional submanifold of a euclidean m-space E^m. Prove that

$$\Delta \tilde{X} = nH,$$

where $\tilde{X}$ is the position vector in E^m.

4. Let M be a closed submanifold of a euclidean m-space E^m. Prove that (i)

$$\Delta(\tilde{X}, \tilde{X}) = 2n(1 + \tilde{g}(H, \tilde{X})),$$

and (ii) M is contained in a small hypersphere of E^m centered at a point $C \in E^m$ if and only if we have either $\tilde{g}(H, \tilde{X} - C) \geqq -1$ or $\tilde{g}(H, \tilde{X} - C) \leqq -1$ everywhere.

5. Give an example to show that, for a submanifold M in a small hypersphere of a euclidean m-space E^m centered at the origin, the normal vector field ξ satisfying the conditions given in Proposition 5.1 is not, in general, unique.

6. Let M be a minimal hypersurface of a euclidean $(n + 1)$-space E^{n+1} which is defined by

$$X^{n+1} = X^{n+1}(X^1, \ldots, X^n)$$

where $X^1, \ldots, X^{n+1}$ are the euclidean coordinates of E^{n+1}.

   (i) Prove that the unit hypersurface normal vector field of M in E^{n+1} has the components

$$\frac{1}{W}\frac{\partial X^{n+1}}{\partial X^1}, \ldots, \frac{1}{W}\frac{\partial X^{n+1}}{\partial X^n}, -\frac{1}{W},$$

where

$$W = \left\{1 + \sum_i \left(\frac{\partial X^{n+1}}{\partial X^i}\right)^2\right\}^{\frac{1}{2}}.$$

   (ii) If $n = 2$, prove that the Gaussian curvature G of M satisfies $G = (W/2)\Delta(1/W)$, where Δ is the Laplacian of M with respect to the induced metric g.

   (iii) Prove that the length of the 1-form $d(W^{-1})$ is given by $|d(W^{-1})|^2 = -G(1 - W^{-2})$, and for any function $\rho(W^{-1})$ of W^{-1}, we have

$$\Delta\rho(W^{-1}) = \rho'(W^{-1})\Delta(W^{-1}) + \rho''(W^{-1})|d(W^{-1})|^2.$$

   (iv) Use (ii) and (iii) to prove that $G = \Delta\log(1 + W^{-1})$.

   (v) Let $g^* = (1 + w^{-1})^2 g$. Prove that the surface M with the metric g^* is a flat surface, that is, the Gaussian curvature of (M, g^*) vanishes.

7. Let M be an n-dimensional submanifold in an m-dimensional space form $R^m(c)$ of curvature c. Prove that if there exists a parallel umbilical section ξ on M, then $M_1(\xi)$ is constant and M is contained in a hypersphere of $R^m(c)$. In particular, if the mean curvature vector is nonzero and parallel and M is pseudoumbilical in $R^m(c)$, then M is a minimal submanifold of a small hypersphere of $R^m(c)$.

8. Let M be a flat surface of a 3-dimensional space form $R^3(c)$ $(c = 0, 1)$ with constant mean curvature. Prove that

   (i) If $c = 0$, then M is an open piece of a circular cylinder, where a circular cylinder means a product surface of a straight line and a plane circle in $E^3 = R^3(0)$.

   (ii) If $c = 1$, then M is an open piece of a product surface of two plane circles.

9. Let M be an n-dimensional submanifold of an m-dimensional Riemannian manifold N and $Y_1, \ldots, Y_n, \xi_1, \ldots, \xi_{m-n}$ be a local field of orthonormal

frame in N such that Y_i are tangent to M and ξ_x normal to M. We put

$$\tilde{\nabla}_X Y_i = \omega_i^j(X)Y_j + \omega_i^{x+n}(X)\xi_x,$$
$$\tilde{\nabla}_X \xi_x = \omega_{x+n}^i(X)Y_i + \omega_{x+n}^{y+n}(X)\xi_y,$$

and let $\omega^1, \ldots, \omega^n$ be the dual vectors of $Y_1, \ldots, Y_n$. Prove that

(i) $\omega_A^B = -\omega_B^A$,

(ii) $\omega_i^{x+n} = h_{ij}^x \omega^j$, where $h_{ij}^x = h^x(Y_i, Y_j)$, and

(iii) *Structure equations*:

$$d\omega^i = -\omega_B^i \wedge \omega^B,$$
$$d\omega_B^A = -\omega_C^A \wedge \omega_B^C + \Phi_B^A,$$
$$\Phi_B^A = \frac{1}{2}\tilde{K}_{CDB}^A \omega^C \wedge \omega^D,$$

where $\tilde{K}_{CDB}^A$ are given by

$$\tilde{\nabla}_{Y_C}\tilde{\nabla}_{Y_D}Z^A - \tilde{\nabla}_{Y_D}\tilde{\nabla}_{Y_C}Z^A - \tilde{\nabla}_{[Y_C, Y_D]}Z^A = K_{CDB}^A Z^B$$

for any vector field $Z = Z^A Y_A$ and $Y_{x+n} = \xi_x$ in N defined along M.

10. Let M be an n-dimensional submanifold of an m-dimensional space form $R^m(c)$ of curvature c and let N_P be the normal subspace at $P \in M$ given by $\{\xi \in T_P^\perp(M): A_\xi \neq 0\} \cup \{0\}$. If N_P has the same dimension everywhere on M, then N_P is said to be invariant under parallelism of the normal bundle if, for any normal vector field η with $\eta_P \in N_P$, $P \in M$, we have $\nabla_X^\perp \eta \in N_P$ for any vector X of M at P. Prove that if the normal subspace N_P is invariant under parallelism of the normal bundle and l is the constant dimension of N_P, then M is contained in a great $(n+l)$-sphere of $R^m(c)$ (Erbacher, 1971).

11. Use the method of Myers (1951) to prove Proposition 5.5.

Chapter 3

Minimal Submanifolds

1 The First Variation

Let M be an n-dimensional manifold immersed in an m-dimensional Riemannian manifold N. Then the submanifold M can be represented locally by

$$u^A = u^A(x^h). \tag{1.1}$$

Let M' be a simply connected region (that is a simply connected open submanifold) in M bounded by a closed submanifold $\partial M'$ in M. We consider the integral

$$I = \int_{M'} L\, dx^1 \wedge dx^2 \wedge \cdots \wedge dx^n, \tag{1.2}$$

where L is a differentiable function of class C^2 of x's and the first derivatives $B_h^A = \partial u^A / \partial x^h$.

Let $f^A(x^h)$ be a set of arbitrary m functions such that

$$f^A = 0 \quad \text{on } \partial M'. \tag{1.3}$$

Then

$$\bar{u}^A = u^A + t f^A \tag{1.4}$$

define another submanifold $\bar{M}$ in N containing ∂M and nearby M, where t is an infinitesimal. When these expressions are substituted for the function L in (1.2), and expanding by Taylor's theorem, we have for $\overline{M}$ the corresponding integral

$$\overline{I} = I + t \int_{M'} \left\{ f^A \frac{\partial L}{\partial u^A} + \frac{\partial f^A}{\partial x^h} \frac{\partial L}{\partial B_h^A} \right\} dx^1 \wedge dx^2 \wedge \cdots \wedge dx^n + o(t^2), \quad (1.5)$$

where $o(t^2)$ is a term involving the second and higher order terms in t. If we put

$$\delta I = t \int_{M'} \left\{ f^A \frac{\partial L}{\partial u^A} + \frac{\partial f^A}{\partial x^h} \frac{\partial L}{\partial B_h^A} \right\} dx^1 \wedge dx^2 \wedge \cdots \wedge dx^n \quad (1.6)$$

and integrate the second term of (1.6) by parts, we obtain

$$\delta I = t \int_{M'} f^A \left\{ \frac{\partial L}{\partial u^A} - \frac{\partial}{\partial x^h} \left(\frac{\partial L}{\partial B_h^A} \right) \right\} dx^1 \wedge dx^2 \wedge \cdots \wedge dx^n, \quad (1.7)$$

by virtue of (1.3). In order that I shall be a minimum for all submanifolds passing through $\partial M'$, it is necessary that $\delta I = 0$ for all sets of functions f^A satisfying (1.3). From (1.7), we see that this condition is

$$\frac{\partial}{\partial x^h} \left(\frac{\partial L}{\partial B_h^A} \right) - \frac{\partial L}{\partial u^A} = 0, \quad B_h^A = \frac{\partial u^A}{\partial x^h}. \quad (1.8)$$

Equation (1.8) is called the *generalized equation of Euler*.

In particular, if the integral (1.2) is the integral of volume, that is,

$$I = \int \sqrt{\mathfrak{g}} dx^1 \wedge dx^2 \wedge \cdots \wedge dx^n, \quad (1.9)$$

then Eq. (1.8) reduces to

$$\frac{\partial}{\partial x^h} \left(\frac{\partial \sqrt{\mathfrak{g}}}{\partial B_h^A} \right) - \frac{\partial \sqrt{\mathfrak{g}}}{\partial u^A} = 0, \quad (1.10)$$

where

$$\mathfrak{g} = \det(g_{ji}) \quad \text{and} \quad g_{ji} = \tilde{g}_{BA} B_j^B B_i^A.$$

Hence, by a direct computation (see, for instance, Eisenhart, 1947), we have

$$0 = \frac{\partial}{\partial x^h} \left(\frac{\partial \sqrt{\mathfrak{g}}}{\partial B_h^A} \right) - \frac{\partial \sqrt{\mathfrak{g}}}{\partial u^A} = \sqrt{\mathfrak{g}} \tilde{g}_{BA} H^B, \quad (1.11)$$

where H^B are the local components of the mean curvature vector H. From (1.11) we obtain the following well-known theorem:

Theorem 1.1. *A minimal submanifold M in a Riemannian manifold N is an extremal for the integral of volume.*

2 Minimal Submanifolds in Euclidean Space

In this section, we shall assume that the ambient space N is a euclidean m-space E^m. The equation of Gauss is then given by

$$K(X, Y; Z, W) = \tilde{g}(h(X, W), h(Y, Z)) - \tilde{g}(h(X, Z), h(Y, W)), \qquad (2.1)$$

or in local components

$$K_{kjih} = h^x_{kh} h_{jix} - h^x_{ki} h_{jhx}, \qquad (2.2)$$

where $h_{jix} = h^x_{ji}$. Transvecting g^{kh} to this equation, we find

$$K_{ji} = -g^{kh} h^x_{ki} h_{jhx} + h^t_{tx} h^x_{ji}, \qquad (2.3)$$

where $h^h_{jx} = h_{jtx} g^{th}$. Thus, if M is a minimal submanifold of E^m, we have

$$K_{ji} = -h^x_{ti} h^t_{jx}. \qquad (2.4)$$

The right-hand side of this equation is negative semidefinite. From (2.4), we see that the scalar curvature is given by

$$r = -h^i_{jx} h^{jx}_i, \qquad (2.5)$$

where $h^{jx}_i = h^j_{ix}$. From (2.4) and (2.5) we have

Theorem 2.1 (Takahashi, 1966). *Let M be a minimal submanifold of a euclidean m-space E^m. Then the Ricci tensor of M is negative semidefinite, and M is totally geodesic if and only if its scalar curvature is zero.*

Let $X^1, \ldots, X^{n+1}$ be the standard coordinates of the euclidean $(n+1)$-space E^{n+1}. We consider a minimal hypersurface M which can be globally represented by an equation of the form

$$X^{n+1} = X^{n+1}(X^1, \ldots, X^n).$$

We call such a hypersurface a *nonparametric hypersurface*.

Theorem 2.2 (Bernstein). *Let*

$$X^3 = X^3(X^1, X^2) \qquad (2.6)$$

be a nonparametric minimal surface in a euclidean 3-space E^3 defined for all X^1 and X^2. Then Eq. (2.6) is a linear function, that is, the minimal surface is a plane in E^3.

Proof. On the minimal surface M we consider the new metric

$$g^* = \left(1 + \frac{1}{W}\right)^2 g, \quad W = \left\{1 + \left(\frac{\partial X^3}{\partial X^1}\right)^2 + \left(\frac{\partial X^3}{\partial X^2}\right)^2\right\}^{\frac{1}{2}}, \quad (2.7)$$

where g is the induced metric of M from the euclidean 3-space E^3. Since the surface M with the induced metric g is complete and

$$g^*(X, Y) \geqq g(X, Y)$$

for any vector fields X, Y on M, the metric g^* is also complete. From Problem 6 of Chapter 2, we know that the Gaussian curvature of M with the new metric g^* is zero, that is, the new metric g^* is a flat metric. Thus M with the metric g^* is parabolic and it is isometric to the euclidean (x, y)-plane with the standard flat metric

$$dx^2 + dy^2.$$

On the other hand, since M is minimal in E^3, the Gaussian curvature of M with the metric tensor g is $\leqq 0$. Since the Laplacian Δ of (M, g) differs from the operator

$$\frac{\partial^2}{\partial x^2} + \frac{\partial^2}{\partial y^2}$$

by a positive factor, we have

$$\left(\frac{\partial^2}{\partial x^2} + \frac{\partial^2}{\partial y^2}\right) \log\left(1 + \frac{1}{W}\right) \leqq 0.$$

Hence, the function $-\log(1 + 1/W)$ considered as a function on M with the metric tensor g^* is a subharmonic function and bounded above by zero. Therefore, $\log(1 + 1/W)$ is a constant. This implies that the Gaussian curvature of M with the metric tensor g is zero. Hence, by Theorem 2.1, we see that M is a plane. This proves the theorem. □

The famous Bernstein problem is to ask whether a nonparametric minimal hypersurface in E^{n+1} with the form

$$X^{n+1} = X^{n+1}(X^1, \ldots, X^n)$$

defined for all $X^1, \ldots, X^n$ is necessarily linear. The answer is known to be affirmative for $n \leqq 7$ [de Giorgi (1965) for $n = 3$, Almgren (1966) for $n = 4$ and Simons (1968) for $n = 5, 6, 7$], and to be negative for $n > 7$ (Bombieri, de Giorgi, and Giusti, 1969) .

3 Minimal Submanifolds of a Submanifold

Let M be an n-dimensional submanifold of a q-dimensional submanifold M' of an m-dimensional Riemannian manifold N.

Let X and Y be any two vector fields on M, and let $\tilde{\nabla}$ and ∇' be the Riemannian connections of N and M', respectively. Then we have

$$\tilde{\nabla}_X Y = \nabla'_X Y + \overline{h}(X, Y), \tag{3.1}$$

where $\overline{h}$ is the second fundamental form of M' in N. Let ∇ be the Riemannian connection of M and h' be the second fundamental form of M in M'. Then we have

$$\nabla'_X Y = \nabla_X Y + h'(X, Y). \tag{3.2}$$

From (3.1) and (3.2), we find

$$\tilde{\nabla}_X Y = \nabla_X Y + h'(X, Y) + \overline{h}(X, Y). \tag{3.3}$$

Thus, the second fundamental form h of M in N is given by

$$h(X, Y) = h'(X, Y) + \overline{h}(X, Y), \tag{3.4}$$

$h'(X, Y)$ being tangent to M' and $\overline{h}(X, Y)$ being normal to M'. Thus, if we denote by H and H' the mean curvature vectors of M in N and M', respectively, then we have

$$H = H' + H(M; M', N), \tag{3.5}$$

where $H(M; M', N)$ is normal to M' and is given by

$$H(M; M', N) = \frac{1}{n} \sum_{i=1}^{n} \overline{h}(E_i, E_i), \tag{3.6}$$

where $E_1, \ldots, E_n$ are orthonormal vector fields in M. We call the normal vector field $H(M; M', N)$, the *relative mean curvature* vector of M with respect to M' and N (Chen and Yano, 1971b). From (3.5) we have

Lemma 3.1. *In order that M be minimal in M', it is necessary and sufficient that the mean curvature vector of M in N be normal to M'.*

Lemma 3.2. *In order that M be minimal in N, it is necessary and sufficient that M be minimal in M' and that the relative mean curvature vector of M with respect to M' and N vanish.*

Let $Z = Z^A \partial_A$ be a vector field in N defined along M. If Z satisfies

$$B_j^A + \tilde{\nabla}_j Z^A = 0, \tag{3.7}$$

then Z is called a *concurrent vector field* along M (Chen and Yano, 1971b).

From (3.7), we can easily prove that

$$H^A + \frac{1}{n} g^{ji} \nabla_j \nabla_i Z^A = 0, \tag{3.8}$$

where H^A are the local components of the mean curvature vector H of M in N. We simply denote (3.8) by

$$H + \frac{1}{n} g^{ji} \nabla_j \nabla_i Z = 0. \tag{3.9}$$

Combining (3.5) and (3.9), we obtain

$$H' + H(M; M', N) + \frac{1}{n} g^{ji} \nabla_j \nabla_i Z = 0. \tag{3.10}$$

Since the relative mean curvature vector $H(M; M', N)$ is normal to M', $H' = 0$ if and only if $\Delta Z^A = g^{ji} \nabla_j \nabla_i Z^A$ is normal to M'. Consequently, we have

Theorem 3.1 (Chen and Yano, 1971b). *Let M be an n-dimensional submanifold of a q-dimensional submanifold M' of an m-dimensional Riemannian manifold N. Suppose that there exists a vector field $Z = Z^A \partial_A$ in N defined along M which is concurrent along M. Then in order for M to be minimal in M', it is necessary and sufficient that $\Delta Z^A = g^{ji} \nabla_j \nabla_i Z^A$ be normal to M.*

If M is a submanifold of a euclidean m-space E^m with $\tilde{X}$ as its position vector with respect to a point in E^m, then it is clear that $-\tilde{X}$ is a concurrent vector field in E^m defined along M. Thus, from Theorem 3.1, we have the following

Corollary 3.1 (Takahashi, 1966). *A submanifold of a euclidean m-space is a minimal submanifold if and only if the position vector field is harmonic.*

Corollary 3.2 (Takahashi, 1966). *Let M be a submanifold of a small hypersphere S^{m-1} of E^m centered at a point C. Then, in order for M to be minimal in S^{m-1}, it is necessary and sufficient that $\Delta \tilde{X} = c\tilde{X}$ for a constant c, where $\tilde{X}$ is the position vector of E^m with respect to the point C.*

Corollary 3.1 follows immediately from Theorem 3.1 and Corollary 3.2 follows immediately from the fact $\Delta \tilde{X} = nH$ and from Theorem 3.1.

Corollary 3.3. *Let M be an n-dimensional submanifold of a euclidean m-space E^m. If the position vector field $\tilde{X}$ of M in E^m with respect to a point $C \in E^m$ is parallel to the mean curvature vector H, then the submanifold M is either a minimal submanifold of E^m or a minimal submanifold of a small hypersphere of E^m centered at the point C.*

Proof. Let

$$\Psi = \{P \in M; H \neq 0 \text{ at } P\}. \tag{3.11}$$

If Ψ is empty, then M is a minimal submanifold of E^m. Now, suppose that Ψ is nonempty. Then, on the open subset Ψ, we may choose a unit normal vector field

η such that η is parallel to the mean curvature vector H. Since the position vector field $\tilde{X}$ of M in E^m with respect to C is parallel to the mean curvature vector H, we may put

$$\tilde{X} = f\eta \tag{3.12}$$

on the open subset Ψ. Let Y be any vector field on M. Then we have

$$\tilde{\nabla}_Y \tilde{X} = (Yf)\eta + f\tilde{\nabla}_Y \eta. \tag{3.13}$$

Since $\tilde{X}$ is the position vector field on E^m and η is unit,

$$\tilde{g}(\tilde{\nabla}_Y \tilde{X}, \eta) = 0, \quad \tilde{g}(\tilde{\nabla}_Y \eta, \eta) = 0.$$

Thus, from (3.13), we obtain $Yf = 0$. This implies that

$$\tilde{g}(\tilde{X}, \tilde{X}) = f^2 = \text{constant}$$

on each component of Ψ.

On the other hand, we have

$$\Delta\tilde{g}(\tilde{X}, \tilde{X}) = 2n(1 + \tilde{g}(\tilde{X}, H)) = 0 \tag{3.14}$$

on Ψ. Hence, by the parallelism of $\tilde{X}$ to H, we see that the mean curvature vector H has constant length on each component of Ψ. Therefore, by the continuity of the mean curvature vector H on M, we know that $\Psi = M$. This implies that M is contained in a small hypersphere S^{m-1} of E^m centered at the point C. Thus, by the identity $\Delta\tilde{X} = nH$, we see that $\Delta\tilde{X}$ is proportional to $\tilde{X}$. In particular, we see that $\Delta\tilde{X}$ is normal to S^{m-1}. From this fact and from Theorem 3.1, we may conclude that M is a minimal submanifold of S^{m-1}. This proves the corollary. $\square$

We now consider a unit vector field $\eta = \eta^A \partial_A$ of N defined along M and normal to M'. We assume that M is umbilical with respect to the normal direction η with $M_1(\eta) = \beta$, where $M_1(\eta) = \frac{1}{n}\text{trace } A_\eta$. From the formula of Gauss for M in N, we find

$$h(X_j, X_i) = \sum_{x=1}^{q-n} h_{jix}\xi_x + \beta g_{ji}\eta_1 + \sum_{x=2}^{m-q} h_{jiz}\eta_x \tag{3.15}$$

where $z = x + q - n$, $\eta_1 = \eta$, and $\xi_1, \ldots, \xi_{q-n}, \eta_1, \ldots, \eta_{m-q}$ are orthonormal normal vector fields of M in N such that $\xi_1, \ldots, \xi_{q-n}$ are tangent to M' and $\eta_1, \ldots, \eta_{m-q}$ are normal to M'. From this we find

$$H = H' + \beta\eta_1 + \frac{1}{n}\sum_{x=2}^{m-q} h_z\eta_x, \tag{3.16}$$

where $h_z = h_{tz}^t$. Thus, from (3.16), we see that, if

$$\tilde{g}(H, H) \leqq \beta^2,$$

then we have $\tilde{g}(H, H) = \beta^2$ and

$$H' = 0, \quad h_{tq-n+2}^t = \cdots = h_{tm-n}^t = 0.$$

This shows that M is minimal in M' and M is minimal in N if and only if $\beta = 0$. Consequently, we have

Proposition 3.1 (**Chen and Yano, 1971b**). *Let η be a unit vector field of N defined along M and normal to M'. Assume that M is umbilical with respect to η with $M_1(\eta) = \beta$. If the mean curvature vector H of M in N satisfies $\tilde{g}(H, H) \leqq \beta^2$, then the submanifold M is minimal in M' and M is minimal in N if and only if $\beta = 0$.*

We now consider the case where M' is totally umbilical in N. In this case, we have

$$\tilde{\nabla}_{X_\kappa} \eta_u = -\alpha_u X_\kappa + l_{\kappa u}^v \eta_v, \tag{3.17}$$

where $X_\kappa = \partial_\kappa$, $\mu, \kappa = 1, \ldots, q$ are the basis vector fields of the manifold M' and $l_{\kappa u}^v = -l_{\kappa v}^u$ the third fundamental tensors of M' in N; where, here and in the sequel, the indices u, v, w run over the range $1, \ldots, m - q$.

From (3.17) we see that the mean curvature vector $\overline{H}$ of M' in N is given by

$$\overline{H} = \frac{1}{q} g^{\kappa\mu} \overline{h}(X_\kappa, X_\mu) = \alpha^u \eta_u \tag{3.18}$$

where $\alpha^u = \alpha_u$, and the relative mean curvature vector $H(M; M', N)$ is given by

$$H(M; M', N) = \frac{1}{n} g^{ji} \overline{h}(X_j, X_i) = \alpha^u \eta_u. \tag{3.19}$$

Consequently, we have

$$\overline{H} = H(M; M', N) = \alpha^u \eta_u \tag{3.20}$$

and

$$H = H' + \overline{H}. \tag{3.21}$$

Therefore, for the covariant derivative of H, we have

$$\tilde{\nabla}_X H = \nabla_X' H' + \overline{h}(X, H') + [-A_{\overline{H}}(X) + D_X \overline{H}]$$
$$= [\nabla_X' H' - \tilde{g}(\overline{H}, \overline{H}) X] + D_X \overline{H} \tag{3.22}$$

where X is any tangent vector field in M. Thus, if the mean curvature vector H of M in N is parallel in the normal bundle, then, from (3.22), we have

$$D_X \overline{H} = 0 \tag{3.23}$$

and consequently, the length of $\overline{H}$ is constant on M.

From (3.23), we find

$$\nabla'_X H' = \tilde{g}(\overline{H}, \overline{H})X \tag{3.24}$$

from which, we find

$$\Delta \overline{H} = g^{ji}\nabla_j \nabla_i \overline{H} = -n\tilde{g}(\overline{H}, \overline{H})\{H' + H(M; M', N)\}. \tag{3.25}$$

Therefore, we obtain the following:

Theorem 3.2 (Chen and Yano, 1971b). *If M' is totally umbilical in N, then the mean curvature vector $\overline{H}$ of M' in N coincides with the relative mean curvature vector of M with respect to M' and N. Moreover, if the mean curvature vector H of M in N is parallel in the normal bundle, then the length of $\overline{H}$ is constant and in the case in which $\overline{H}$ is different from zero, in order for M to be minimal in M', it is necessary and sufficient that*

$$\Delta \overline{H} = -n\tilde{g}(\overline{H}, \overline{H})H(M; M', N)$$

on M.

From Theorem 3.2 we have immediately the following:

Corollary 3.4 (Chen and Yano, 1971b). *Let M' be a totally umbilical hypersurface in N with nonzero constant mean curvature or be totally umbilical in N of constant curvature. Then, in order that M is minimal in M', it is necessary and sufficient that the unit normal vector field C of M' in N satisfy $\Delta C = g^{ji}\nabla_j \nabla_i C = fC$ for a constant f.*

It is obvious that this corollary is a generalization of Corollaries 3.1 and 3.2 of Takahashi.

4 Examples of Minimal Submanifolds

In this section, we shall give some examples of minimal submanifolds.

Example 4.1. Every totally geodesic submanifold of a Riemannian manifold is a minimal submanifold. In particular, every great hypersphere of a space form $R^m(k)$ is a minimal submanifold.

Example 4.2. Let M be a surface in a euclidean 3-space E^3 given by

$$X^1 = r\cos\varphi, \quad X^2 = r\sin\varphi,$$

$$X^3 = a\cosh^{-1}\frac{r}{a}, \quad a = \text{constant} \neq 0.$$

Then M is a minimal surface in E^3. We call the surface M a *catenoid*.

Example 4.3. Let M be a surface in E^3 given by

$$X^1 = r\cos\varphi, \quad X^2 = r\sin\varphi, \quad X^3 = a\varphi, \quad a = \text{constant} \neq 0.$$

Then M is a minimal surface in E^3. We call the surface M a *right helicoid*.

Example 4.4. Let $S^q(r)$ denote a q-dimensional sphere in E^{q+1} with radius r. Let n and p be two positive integers such that $p < n$ and $M_{p,n-p}$ the product manifold given by

$$M_{p,n-p} = S^p\left(\sqrt{\frac{p}{n}}\right) \times S^{n-p}\left(\sqrt{\frac{n-p}{n}}\right).$$

We imbed $M_{p,n-p}$ into $S^{n+1}(1)$ as follows. Let (X_1, X_2) be a point of $M_{p,n-p}$ where X_1 (resp. X_2) is a vector in E^{p+1} (resp. E^{n-p+1}) of length $\sqrt{p/n}$ (resp. $\sqrt{(n-p)/n}$). We consider (X_1, X_2) as a unit vector in

$$E^{n+2} = E^{p+1} \times E^{n-p+1}.$$

Then $M_{p,n-p}$ is a minimal submanifold of $S^{n+1}(1)$. We call this minimal hypersurface of $S^{n+1}(1)$ a *Clifford minimal hypersurface*. In particular, if $n = 2$, and $p = 1$, $M_{1,1}$ is a flat minimal surface of $S^3(1)$. We call this minimal surface the *Clifford torus*.

Example 4.5. Let (x, y, z) be the standard coordinate system in E^3 and $(u^1, u^2, u^3, u^4, u^5)$ be the standard coordinate system in E^5. We consider the mapping defined by

$$u^1 = \frac{1}{\sqrt{3}}yz, \quad u^2 = \frac{1}{\sqrt{3}}zx, \quad u^3 = \frac{1}{\sqrt{3}}xy,$$

$$u^4 = \frac{1}{2\sqrt{3}}(x^2 - y^2), \quad u^5 = \frac{1}{6}(x^2 + y^2 - 2z^2).$$

This defines an isometric immersion of $S^2(\sqrt{3})$ into $S^4(1)$. Two points (x, y, z) and $(-x, -y, -z)$ of $S^2(\sqrt{3})$ are mapped into the same point of $S^4(1)$, and this mapping defines an imbedding of the real projective plane into $S^4(1)$. This real projective plane imbedded in $S^4(1)$ is called the *Veronese surface*. It is a minimal surface of $S^4(1)$.

Example 4.6. H.B. Lawson has constructed closed orientable minimal surface of arbitrary genus in $S^3(1)$ (Lawson, 1970).

Example 4.7. Let n and s be two positive integers and let $k(s)$ and $m(s)$ be the real numbers given by

$$k(s) = \frac{n}{s(s+n-1)},$$

$$m(s) = (2s+n-1)\frac{(n+s-2)!}{s!(n-1)!} - 1.$$

M. do Carmo and N.R. Wallach have constructed an isometric minimal immersion $\Psi_{n,s}$ from an n-sphere of constant curvature $k(s)$ into the unit $m(s)$-sphere $S^{m(s)}(1)$. Moreover, they showed that any isometric minimal immersion φ of an n-sphere into a unit hypersphere $S^{m(s)}(1)$ of $E^{m(s)+1}$ must be a product immersion of $\Psi_{n,s}$ and a symmetric, positive semidefinite linear map A of $E^{m(s)+1}$, that is, φ is equivalent to $A \circ \Psi_{n,s}$. For $n = 2$, the map A is the identity map (Calabi, 1967).

Example 4.8. Every complex submanifold of a Kaehler manifold is minimal.

5 Inequality of Simons

Let M be an n-dimensional submanifold of an m-dimensional Riemannian manifold N of constant curvature c. Let h be the second fundamental form of M in N. Then the Laplacian of the length of h, $\langle h \rangle$, satisfies

$$\frac{1}{2}\Delta\langle h\rangle^2 = g^{kl}(\overline{\nabla}_k\overline{\nabla}_l h^x_{ji})h^{ji}_x + (\overline{\nabla}_k h^x_{ji})(\overline{\nabla}^k h^{ji}_x), \tag{5.1}$$

where $\overline{\nabla}$ is the connection of van der Wearder-Bortolotti and $\overline{\nabla}^k = g^{kt}\overline{\nabla}_t$.

By using (II.6.9) and (II.6.11), we find

$$\frac{1}{2}\Delta\langle h\rangle^2 = g^{kl}\{\overline{\nabla}_j\overline{\nabla}_k h^x_{il} - K^t_{kji}h^x_{tl} - K^t_{kjl}h^x_{it} + K^x_{kjy}h^y_{il}\}h^{ji}_x + (\overline{\nabla}_k h^x_{ji})(\overline{\nabla}^k h^{ji}_x)$$

$$= n(\overline{\nabla}_j\overline{\nabla}_i h^{(x)})h^{ji}_x + K^t_j h^x_{it}h^{ji}_x - K_{kjit}h^{kt}_x h^{jix} + K^x_{kjy}h^{ky}_i h^{ji}_x$$

$$+ (\overline{\nabla}_k h^x_{ji})(\overline{\nabla}^k h^{ji}_x),$$

where $h^{jix} = h^{ji}_x$ and $K^h_j = g^{th}K_{jt}$.

If we substitute (II.6.10), (II.6.13), and (II.6.14) into the preceding equation, we find

$$\frac{1}{2}\Delta\langle h\rangle^2 = n(\overline{\nabla}_j\overline{\nabla}_i h^{(x)})h^{ji}_x + \{(n-1)cg_{jt} + nh^{(y)}h_{jty} - h^y_{jk}h^k_{ty}\}h^t_{ix}h^{jix}$$

$$- \{c(g_{kt}g_{ji} - g_{jt}g_{ki}) + (h^y_{kt}h_{jiy} - h^y_{jt}h_{kiy})\}h^{kt}_x h^{jix}$$

$$+ \{h^t_{ky}h^x_{jt} - h^t_{jy}h^x_{kt}\}h^{ky}_i h^{ji}_x + (\overline{\nabla}_k h^x_{ji})(\overline{\nabla}^k h^{ji}_x).$$

Therefore, we have

$$\frac{1}{2}\Delta\langle h\rangle^2 = n(\overline{\nabla}_j\overline{\nabla}_i h^{(x)})h^{ji}_x + nc\langle h\rangle^2 - n^2 c|H|^2$$

$$+ \{nh^x_{ji}h^j_{ky}h^{ik}h^{(y)} - h^x_{ji}h^{ji}_y h^x_{kl}h^{kl}_y\}$$

$$- \{h^x_{ik}h^k_{jy} - h^y_{ik}h^k_{jx}\}\{h^x_y h^{jy}_l - h^{il}_y h^{jy}_l\} + (\overline{\nabla}_k h^x_{ji})(\overline{\nabla}^k h^{ji}_x). \tag{5.2}$$

For a tensor A of type $(1,1)$, we define

$$N(A) = \text{trace } A \circ {}^t A = |A|^2, \tag{5.3}$$

then, (5.2) can be rewritten as follows:

$$\frac{1}{2}\Delta \langle h \rangle^2 = n(\overline{\nabla}_j \overline{\nabla}_i h^{(x)}) h_x^{ji} + nc\langle h \rangle^2 - n^2 c |H|^2 - \sum_{x,y} (\text{trace}(A_x A_y))^2$$

$$- \sum_{x,y} N(A_x A_y - A_y A_x) + \sum_{x,y} (nh^{(y)}) \text{trace}(A_x A_y A_x)$$

$$+ (\overline{\nabla}_k h_{ji}^x)(\overline{\nabla}^k h_x^{ji}). \tag{5.4}$$

In particular, if M is minimal in N, then $h^{(x)} = 0$. Hence, (5.4) reduces to

$$\frac{1}{2}\Delta \langle h \rangle^2 = nc\langle h \rangle^2 - \sum_x (N(A_x))^2 - \sum_{x,y} N(A_x A_y - A_y A_x)$$

$$+ (\overline{\nabla}_k h_{ji}^x)(\overline{\nabla}^k h_x^{ji}). \tag{5.5}$$

We need the following lemma.

Lemma 5.1 (Chern, do Carmo, and Kobayashi, 1970). *Let A and B be two symmetric $(n \times n)$-matrices. Then*

$$N(AB - BA) \leqq 2N(A) \cdot N(B),$$

and the equality holds for nonzero matrices A and B if and only if A and B can be transformed simultaneously by an orthogonal matrix into scalar multiples of A' and B', respectively, where

$$A' = \begin{pmatrix} 0 & 1 & 0 \\ 1 & 0 & \\ \hline & 0 & 0 \end{pmatrix}; \quad B' = \begin{pmatrix} 1 & 0 & 0 \\ 0 & -1 & \\ \hline & 0 & 0 \end{pmatrix}.$$

Moreover, if A_1, A_2, and A_3 are three $(n \times n)$-matrices and if

$$N(A_\alpha A_\beta - A_\beta A_\alpha) = 2N(A_\alpha) \cdot N(A_\beta), \quad 1 \leqq \beta, \alpha \leqq 3,$$

then at least one of the matrices A_α must be zero.

Proof. We may assume that B is diagonal and we denote by $b_1, \ldots, b_n$ the diagonal entries in B. By a simple calculation, we obtain

$$N(AB - BA) = \sum_{i \neq k} (a_{ik})^2 (b_i - b_k)^2,$$

where $A = (a_{ij})$. Since $(b_i - b_k)^2 \leq 2(b_i^2 + b_k^2)$, we obtain

$$
\begin{aligned}
N(AB - BA) &= \sum_{i \neq k} (a_{ik})^2 (b_i - b_k)^2 \\
&\leq 2 \sum_{i \neq k} (a_{ik})^2 (b_i^2 + b_k^2) \\
&\leq 2 \left(\sum_{i,k} (a_{ik})^2 \right) \left(\sum_j b_j^2 \right) \\
&= 2N(A)N(B).
\end{aligned}
\tag{5.6}
$$

Now, we assume that A and B are nonzero matrices and that the equality holds. Then the equality must hold everywhere in (5.6). From the second equality in (5.6), it follows that

$$
a_{11} = \cdots = a_{nn} = 0,
$$

and that

$$
b_i + b_j = 0 \text{ if } a_{ij} \neq 0.
$$

Without loss of generality, we may assume that $a_{12} \neq 0$. Then $b_1 = -b_2$. From the third equality, we now obtain

$$
b_3 = \cdots = b_n = 0.
$$

Since $B \neq 0$, we must have $b_1 = -b_2 \neq 0$, and we conclude that $a_{ik} = 0$ for $(i, k) \neq (1, 2)$. To prove the last statement, let A_1, A_2, A_3 be all nonzero symmetric matrices. From the second statement we have just proved, we see that one of these matrices can be transformed to a scalar multiple of A' as well as to a scalar multiple of B' by orthogonal matrices. But this is impossible since A' and B' are not orthogonally equivalent. This proves the lemma. □

Applying Lemma 5.1 to (5.5), we obtain

$$
\begin{aligned}
-g^{kl} (\overline{\nabla}_k \overline{\nabla}_l h_{ji}^x) h_x^{ji} &= (\overline{\nabla}_k h_{ji}^x)(\overline{\nabla}^k h_x^{ji}) - \frac{1}{2} \Delta \langle h \rangle^2 \\
&\leq 2 \sum_{x \neq y} N(A_x)N(A_y) + \sum_x (N(A_x))^2 - nc \langle h \rangle^2 \\
&= \left(\sum_x N(A_x) \right)^2 + 2 \sum_{x > y} N(A_x)N(A_y) - nc \langle h \rangle^2 \\
&= (m-n)^2 (\sigma_1)^2 + (m-n)(m-n-1)\sigma_2 - nc \langle h \rangle^2, \tag{5.7}
\end{aligned}
$$

where

$$
(m-n)\sigma_1 = \sum_x N(A_x) = \langle h \rangle^2,
$$

and

$$\frac{(m-n)(m-n-1)}{2}\sigma_2 = \sum_{x<y} N(A_x)N(A_y).$$

It can be easily verified that

$$(m-n)^2(m-n-1)(\sigma_1^2 - \sigma_2) = \sum_{x<y}(N(A_x) - N(A_y))^2 \geqq 0. \qquad (5.8)$$

Hence, (5.7) reduces to

$$-g^{kl}(\overline{\nabla}_k\overline{\nabla}_l h_{ji}^x)h_x^{ji} \leqq (m-n)^2\sigma_1^2 + (m-n)(m-n-1)\sigma_2 - nc\langle h\rangle^2$$

$$= (2(m-n)^2 - (m-n))\sigma_1^2 - (m-n)(m-n-1)(\sigma_1^2 - \sigma_2)$$

$$- nc\langle h\rangle^2$$

$$\leqq (m-n)(2(m-n) - 1)\sigma_1^2 - nc\langle h\rangle^2$$

$$= \left(2 - \frac{1}{m-n}\right)\langle h\rangle^4 - nc\langle h\rangle^2. \qquad (5.9)$$

From (5.1), we find that if M is orientable and closed, then we have

$$\int_M g^{kl}(\overline{\nabla}_k\overline{\nabla}_l h_{ji}^x)h_x^{ji}dV = -\int_M (\overline{\nabla}_k h_{ji}^x)(\overline{\nabla}^k h_x^{ji})dV$$

$$\leqq 0. \qquad (5.10)$$

Therefore, from (5.9) and (5.10), we obtain the following inequality of Simons.

Theorem 5.1 (Simons, 1968). *Let M be an n-dimensional oriented closed minimal submanifold of an m-dimensional Riemannian manifold of constant curvature c. Then we have*

$$\int_M \left\{\left(2 - \frac{1}{m-n}\right)\langle h\rangle^2 - nc\right\}\langle h\rangle^2 dV \geqq 0. \qquad (5.11)$$

From Theorem 5.1, we have immediately the following:

Theorem 5.2 (Simons, 1968). *Let M be an n-dimensional closed minimal submanifold of a unit m-sphere. If M is not totally geodesic and if*

$$\langle h\rangle^2 \leqq n \Big/ \left(2 - \frac{1}{m-n}\right)$$

everywhere on M, then $\langle h\rangle^2 = n/(2 - \frac{1}{m-n})$.

As a consequence of Theorem 5.2, it would be of interest to study the minimal submanifolds of S^m with

$$\langle h\rangle^2 = n \Big/ \left(2 - \frac{1}{m-n}\right).$$

Next to the great spheres these can be considered as the "simplest" minimal submanifolds. Since $\langle h \rangle^2$ is constant in this case, (5.1) implies

$$g^{kl}(\overline{\nabla}_k \overline{\nabla}_l h^x_{ji}) h^{ji}_x + (\overline{\nabla}_k h^x_{ji})(\overline{\nabla}^k h^{ji}_x) = 0.$$

Combining this with (5.9), we find

$$0 = \left\{ \left(2 - \frac{1}{m-n}\right) \langle h \rangle^2 - nc \right\} \langle h \rangle^2$$
$$\geq (\overline{\nabla}_k h^x_{ji})(\overline{\nabla}^k h^{ji}_x).$$

This implies that $\overline{\nabla}_k h^x_{ji} = 0$, that is, the second fundamental form h is covariant constant with respect to the van der Waerden-Bortolotti connection. By using this fact and applying Lemma 5.1, S.S. Chern, M. do Carmo, and S. Kobayashi prove the following:

Theorem 5.3 (Chern, do Carmo, and Kobayashi, 1970). *The open pieces of the Clifford minimal hypersurfaces and the Veronese surface are the only minimal. submanifolds on a unit m-sphere with*

$$\langle h \rangle^2 = n \bigg/ \left(2 - \frac{1}{m-n}\right).$$

This theorem is a local theorem. For the proof, see the paper of Chern, do Carmo, and Kobayashi (1970).

Problems

1. Prove formula (1.11).

2. Prove that Examples 4.2, 4.3, 4.4, 4.5, and 4.8 are minimal submanifolds.

3. Prove Eq. (5.1).

4. Prove that if M is a minimal surface with constant Gaussian curvature in a 3-dimensional space form $R^3(c)$ of curvature c, then either M is totally geodesic or $c > 0$ and M is an open piece of a Clifford torus (Chen, 1972d; Lawson, 1969).

5. Let M be a complete minimal surface in a 3-sphere S^3. Prove that:

   (i) If the Gaussian curvature $G \leq 0$ everywhere, then

   $$\log \sqrt{1-G}$$

   is a superharmonic function bounded from below on M.

(ii) If the Gaussian curvature $G \geq 0$ everywhere, then G is a superharmonic function on M.

(iii) Use (i) and (ii) to prove that every complete minimal surface in S^3 with Gaussian curvature ≥ 0 or ≤ 0 everywhere is either a clifford torus or a great sphere in S^3 (Itoh, 1970).

[Hint: If the Gaussian curvature of M is nonnegative everywhere and M is complete noncompact without boundary, then M is parabolic (Huber, 1957)].

6. Let $x: S^2 \to S^4(1)$ be an immersion of a 2-sphere into a unit 4-sphere as a minimal surface. Prove that if the Euler number of the normal bundle $T^\perp(S^2)$ of S^2 in $S^4(1)$ vanishes, then the immersion x is totally geodesic (Ruh, 1971).

7. Let M be a hypersurface of an $(n+1)$-dimensional space form $R^{n+1}(k)$ of curvature k. Suppose that the second fundamental form h of M in $R^{n+1}(k)$ is a covariant constant with respect to the connection of van der Wearden-Bortolotti. Prove that

   (i) There are only two distinct principal curvatures, say λ_1 and λ_2, such that $\lambda_1 \lambda_2 = -k$ and λ_1 and λ_2 are constants.

   (ii) M is locally isometric to the product manifold of two manifolds of constant curvature $k + (\lambda_1)^2$ and $k + (\lambda_2)^2$.

   (iii) If $R^{n+1}(k)$ is a unit $(n+1)$-sphere, then, up to isometries of $R^{n+1}(k)$, M is an open piece of a Clifford minimal hypersurface.

   (iv) If $R^{n+1}(k)$ is euclidean, then, up to rigid motions of E^{n+1}, M is an open piece of the product submanifold of a small s-sphere of E^{n+1} and a great $(n-s)$-sphere of E^{n+1} (Lawson, 1969).

8. Prove Theorem 5.3 of Chern, do Carmo, and Kobayashi.

Chapter 4

Submanifolds with Parallel Mean Curvature Vector

1 Flat Normal Connection

Let M be an n-dimensional submanifold of an m-dimensional Riemannian manifold N. Let $\nabla^\perp$ denote the induced connection in the normal bundle $T^\perp(M)$. If the curvature tensor K^N of the normal connection $\nabla^\perp$ vanishes, that is,

$$K^N(X, Y) = \nabla_X^\perp \nabla_Y^\perp - \nabla_Y^\perp \nabla_X^\perp - \nabla_{[X,Y]}^\perp = 0 \qquad (1.1)$$

for any vector fields X, Y in M, then the normal connection $\nabla^\perp$ is said to be *flat*.

Proposition 1.1. *Let M be an n-dimensional submanifold of an m-dimensional Riemannian manifold N. Then the normal connection $\nabla^\perp$ of M in N is flat if and only if there exist locally $m - n$ mutually orthogonal unit normal vector fields ξ_x such that each of the ξ_x is parallel in the normal bundle.*

Proof. Suppose that there exist locally $m - n$ mutually orthogonal unit normal vector fields ξ_x such that each of ξ_x is parallel in the normal bundle. Then we have

$$\nabla_X^\perp \xi_x = 0 \qquad (1.2)$$

for any vector field X in M. Hence, we have

$$K^N(X, Y)\xi_x = 0. \qquad (1.3)$$

Thus, for any normal vector field $\xi = f^* \xi_x$ of M, we have

$$K^N(X, Y)\xi = \{(XY - YX)f^*\}\xi_x - ([X, Y]f^*)\xi_x = 0. \qquad (1.4)$$

This shows that the normal connection $\nabla^\perp$ is flat.

Conversely, suppose that the normal connection $\nabla^\perp$ is flat. Then, from (1.1), we have

$$\nabla_X^\perp \nabla_Y^\perp \xi_x - \nabla_Y^\perp \nabla_X^\perp \xi_x - \nabla_{[X,Y]}^\perp \xi_x = 0, \qquad (1.5)$$

for any $m - n$ mutually orthogonal unit normal vector fields ξ_x of M. Thus, if we put

$$\nabla_X^\perp \xi_x = \omega_x^y(X)\xi_y, \qquad (1.6)$$

then we have $\omega_x^y = -\omega_y^x$ and

$$0 = \{X\omega_x^y(Y) - Y\omega_x^y(X) - \omega_x^y([X, Y])\}\xi_y + \{\omega_x^z(Y)\omega_z^y(X) - \omega_x^z(X)\omega_z^y(Y)\}\xi_y,$$

that is,

$$d\omega_x^y = -\omega_x^z \wedge \omega_z^y, \quad \omega_x^y = -\omega_y^x. \qquad (1.7)$$

Thus, from Problem 9 of Chapter 1, we know that there exists an $(m - n) \times (m - n)$-matrix $A = (a_x^y)$ of functions satisfying

$$dA = -A\Omega, \quad A^T = A^{-1}, \qquad (1.8)$$

where $\Omega = (\omega_x^y)$. Equation (1.8) has the local expression

$$da_x^z = \sum a_x^y \omega_z^y = -\sum a_x^y \omega_y^z. \qquad (1.9)$$

Put

$$\xi_x' = a_x^y \xi_y. \qquad (1.10)$$

The ξ_x' are also $m - n$ mutually orthogonal unit normal vector fields of M. From (1.9) and (1.10), we find that

$$\omega_x'^y a_y^z = da_x^z + a_x^y \omega_y^z = 0, \qquad (1.11)$$

where $\omega_x'^y$ are defined by $\nabla^\perp \xi_x' = \omega_x'^y \xi_y'$. This shows that each of the ξ_x' is parallel in the normal bundle. This proves the proposition. $\qquad \square$

If the ambient space N is of constant curvature, then we have

Proposition 1.2 (Cartan, 1946). *Let M be an n-dimensional submanifold of a space N of constant curvature. Then the normal connection $\nabla^\perp$ is flat if and only if all of the second fundamental tensors A_x are simultaneously diagonalizable.*

This proposition follows immediately from the equation (II.2.16) of Ricci.

Proposition 1.3. *Let M be an n-dimensional submanifold of an m-dimensional Riemannian manifold N. Suppose that ξ_x and ξ_x' are two sets of $m - n$ mutually*

orthogonal unit normal vector fields of M such that each of the ξ_x and ξ'_x is parallel in the normal bundle. If we have

$$\xi'_x = a^y_x \xi_y \tag{1.12}$$

on a component of the intersection of the domains of definition, then the a^y_x are all constants. In particular, if each $\xi_x = \xi'_x$ at a point, then $(a^y_x) = (\delta^y_x)$.

This proposition follows immediately from the definition of parallelism of a normal vector field in the normal bundle.

2 Surfaces with Parallel Mean Curvature Vector

In this section, we shall assume that M is a surface in an m-dimensional Riemannian manifold N of constant curvature c. By passing, if necessary, to the twofold covering surface, we may assume that M is orientable. It is then possible to choose a system of isothermal coordinates $\{x^1, x^2\}$ covering M (see, for instance, Chern, 1955b). The induced metric tensor g then has the following form:

$$g = E\{(dx^1)^2 + (dx^2)^2\}. \tag{2.1}$$

In the following, we put $X_i = \partial/\partial x^i$ and

$$L = h(X_1, X_1), \quad M = h(X_1, X_2), \quad N = h(X_2, X_2). \tag{2.2}$$

For the isothermal coordinates x^1, x^2 with the metric tensor given by (2.1), we have

$$g^{ij} = \frac{\delta_{ij}}{E}, \quad \Gamma^1_{11} = \Gamma^2_{12} = -\Gamma^1_{22} = \frac{X_1 E}{2E},$$

$$\Gamma^2_{22} = \Gamma^1_{12} = -\Gamma^2_{11} = \frac{X_2 E}{2E}, \tag{2.3}$$

where Γ^h_{ji} are given by

$$\nabla_{X_j} X_i = \Gamma^h_{ji} X_h. \tag{2.4}$$

Therefore, the equation (II.2.18) of Codazzi gives

$$\nabla^{\perp}_{X_2} L - \nabla^{\perp}_{X_1} M = (X_2 E) H, \tag{2.5}$$

$$\nabla^{\perp}_{X_2} M - \nabla^{\perp}_{X_1} N = -(X_1 E) H. \tag{2.6}$$

On the other hand, by the definition of the mean curvature vector H, we have

$$(X_i E) H = -E \nabla^{\perp}_{X_i} H + \frac{1}{2}(\nabla^{\perp}_{X_i} L + \nabla^{\perp}_{X_i} N). \tag{2.7}$$

Therefore, by substituting (2.7) into (2.5) and (2.6), we find

$$\nabla^{\perp}_{X_2}\left(\frac{L-N}{2}\right) - \nabla^{\perp}_{X_1}M = -E\nabla^{\perp}_{X_2}H, \qquad (2.8)$$

$$\nabla^{\perp}_{X_1}\left(\frac{L-N}{2}\right) + \nabla^{\perp}_{X_2}M = E\nabla^{\perp}_{X_1}H. \qquad (2.9)$$

Hence, we have the following:

Lemma 2.1. *Let M be a surface in an m-dimensional Riemannian manifold of constant curvature. Then the mean curvature vector H is parallel in the normal bundle if and only if we have*

$$\nabla^{\perp}_{X_2}\left(\frac{L-N}{2}\right) - \nabla^{\perp}_{X_1}M = \nabla^{\perp}_{X_1}\left(\frac{L-N}{2}\right) + \nabla^{\perp}_{X_2}M = 0$$

for any set of isothermal coordinates x^1, x^2 in M.

Since the normal connection $\nabla^{\perp}$ is metric, we have

Lemma 2.2. *Let M be a surface in an m-dimensional Riemannian manifold of constant curvature. If there exists a parallel isoperimetric section ξ on M, then the function*

$$\varphi(\xi) = \left\langle\frac{L-N}{2},\xi\right\rangle - \langle M,\xi\rangle i \qquad (2.10)$$

is analytic in $z = x^1 + ix^2$ for every set of isothermal coordinates x^1, x^2, where an isoperimetric section ξ on M means a unit normal vector field defined globally on M with $M_1(\xi) = constant$ and $\langle\,,\,\rangle$ stands for $\tilde{g}(\,,\,)$. In particular, if the mean curvature vector H is nonzero and parallel, then the function

$$\varphi\left(\frac{H}{|H|}\right) = \left\langle\frac{L-N}{2},\frac{H}{|H|}\right\rangle - \left\langle M,\frac{H}{|H|}\right\rangle i \qquad (2.11)$$

is analytic in $z = x^1 + ix^2$.

These two lemmas follows immediately from (2.8) and (2.9).

Now, we assume that η is a parallel isoperimetric section on M. Then, by using the function $\varphi(\eta)$ defined by (2.10), we may define an analytic quadratic differential Φ on M which is given locally by

$$\Phi = \varphi(\eta)dz^2, \quad z = x^1 + ix^2 \qquad (2.12)$$

on every domain of the isothermal coordinates x^1, x^2. It is easy to verify that Φ is well-defined on M.

If the codimension is two and the mean curvature vector H has constant length, then the unit normal vector field η' orthogonal to η is also a parallel isoperimetric section on M. Hence the function

$$\varphi(\eta') = \left\langle \frac{L-N}{2}, \eta' \right\rangle - \langle M, \eta' \rangle i$$

is also analytic in $z = x^1 + i x^2$. Similarly, we may define an analytic quadratic differential Φ' on M which is given locally by

$$\Phi' = \varphi(\eta') dz^2. \qquad (2.13)$$

Let $\overline{\varphi}$ denote the conjugate of φ. Then we have

$$\mathrm{Im}(\varphi(\eta')\overline{\varphi}(\eta)) = (\mathrm{Re}\,\overline{\varphi}(\eta))(\mathrm{Im}\,\varphi(\eta')) + (\mathrm{Im}\,\overline{\varphi}(\eta))(\mathrm{Re}\,\varphi(\eta')). \qquad (2.14)$$

The right-hand side of (2.14) vanishes by the parallelism of η and η' in the normal bundle by virtue of Proposition 1.1. Hence, if $\Phi \not\equiv 0$, then

$$\beta = \Phi'/\Phi = \varphi(\eta')/\varphi(\eta) = \varphi(\eta')\overline{\varphi}(\eta)/\varphi(\eta)\overline{\varphi}(\eta) \qquad (2.15)$$

is a real function on M with only isolated poles. On the other hand, since $\varphi(\eta)$ and $\varphi(\eta')$ are both analytic, β is meromorphic. Therefore Φ'/Φ is a real constant unless Φ is identically zero. If $\Phi \equiv 0$, then $A_\eta = \frac{1}{2}(\text{trace } A_\eta)I$ with $\nabla^\perp \eta = 0$. If $\Phi \not\equiv 0$, we may put $\beta = -\tan\alpha$, then the vector field

$$\overline{\eta} = (\sin\alpha)\eta + (\cos\alpha)\eta'$$

is a parallel isoperimetric section on M with

$$A_{\overline{\eta}} = \lambda I, \quad 2\lambda = (\sin\alpha)(\text{trace } A_\eta) + (\cos\alpha)(\text{trace } A_{\eta'}).$$

Hence, by applying Problem II.7, we have the following

Lemma 2.3 (Chen, 1973c; Yau, 1974). *Let M be a surface in a 4-dimensional space form $R^4(c)$ with constant mean curvature $|H|$. If there exists a parallel isoperimetric section η on M, then M is contained in a hypersphere of $R^4(c)$, that is, M is either contained in a great hypersphere of $R^4(c)$ or contained in a small hypersphere of $R^4(c)$.*

Now we are going to prove the following classification theorem:

Theorem 2.1 (Chen, 1973c; Yau, 1974). *Let M be a surface in an m-dimensional space form $R^m(c)$ of curvature c. If the mean curvature vector H is parallel in the normal bundle, then M is one of the following surfaces:*

(i) minimal surfaces of $R^m(c)$,

(ii) minimal surfaces of a small hypersphere of $R^m(c)$, or

(iii) surfaces with constant mean curvature $|H|$ in a 3-sphere of $R^m(c)$.

Proof. We first prove the following lemmas.

Lemma 2.4. *Let M be a surface in an m-dimensional space form $R^m(c)$ of curvature c. If the mean curvature vector H is parallel in the normal bundle and M is neither a minimal surface of $R^m(c)$ nor a minimal surface of a small hypersphere of $R^m(c)$, then the normal connection $\nabla^\perp$ of M in $R^m(c)$ is flat.*

Proof. Since the ambient space is a space form, from the equation (II.2.16) of Ricci, we have

$$K^n(X, Y; \xi, \eta) = g([A_\xi, A_\eta](X), Y). \tag{2.16}$$

Since the mean curvature vector H is nonzero and parallel, H has nonzero constant length. Hence, we may choose $m - n$ mutually orthogonal unit normal vector fields ξ_x such that $H = |H|\xi_1$. In this case, we have

$$\nabla^\perp_X \xi_1 = \omega^x_1 \xi_x = 0. \tag{2.17}$$

Hence, from (2.16), we find

$$[A_1, A_x] = 0. \tag{2.18}$$

On the other hand, since $H = |H|\xi_1$, we have

$$\text{trace } A_x = 0, \quad \text{for } x = 2, \ldots, m - 2. \tag{2.19}$$

Combining (2.18) and (2.19), we see that if A_1 is not proportional to the identity transformation at a point $P \in M$, then we have $[A_x, A_y] = 0$ at the point P for all x and y.

On the other hand, from Lemma 2.2, we know that if the analytic quadratic differential $\Psi = \varphi(H/|H|)dz^2$ is not identically zero, then Ψ has only isolated zeros. From this fact and the definition of the analytic function $\varphi(H/|H|)$, we see that if Ψ is not identically zero, then $[A_x, A_y] = 0$ everywhere on M for all x, y, that is, if Ψ is not identically zero, then the curvature tensor K^N vanishes, that is, the normal connection $\nabla^\perp$ is flat. If Ψ is identically zero, then, by the definition of Ψ, we know that M is pseudoumbilical in $R^m(c)$. Thus, by applying Problem II.7, we see that M is a minimal submanifold of a small hypersphere of $R^m(c)$. This proves the lemma. $\square$

Lemma 2.5. *Let M be a surface in an m-dimensional space form $R^m(c)$ of curvature c. If the mean curvature vector H is parallel in the normal bundle and the normal connection $\nabla^\perp$ is flat, then M is contained in a great 4-sphere of $R^m(c)$.*

Proof. Since the mean curvature vector H is parallel in the normal bundle, H has constant length. In the following, let N_P be the normal subspace of M in $R^m(c)$ given by

$$N_P = \{\xi \in T^\perp_P(M); \tilde{g}(H, \xi) = 0\}. \tag{2.20}$$

We define a linear mapping ρ from N_P into the set of all symmetric matrices of order 2, $\mathfrak{S}_2$, by

$$\rho(\xi) = A_\xi. \tag{2.21}$$

Let $0_P = \rho^{-1}(0)$ and N'_P be the normal subspace of N_P given by

$$N_P = N'_P \oplus 0_P, \quad N'_P \perp 0_P.$$

Then, by the assumption $K^N = 0$, we see, from Proposition 1.2, that

$$\dim N'_P \leqq 1. \tag{2.22}$$

In the following, let $M_0 = \{P \in M: \dim N'_P = 0\}$ and $M_1 = \{P \in M: \dim N'_P = 1\}$. It is clear that M_1 is an open subset of M. We consider the cases $H = 0$ and $H \neq 0$, separately.

Case (i) $H = 0$. In this case, M_0 is the subset of M where the second fundamental form h vanishes. On the subset M_1 we may choose a unit normal vector field ξ_1 such that $\xi_1 \in N'_P$ for every point $P \in M_1$. Hence, for any local field of orthonormal frames ξ_x with $\xi_1 \in N'_P$, we have

$$A_x = 0, \quad \text{for } x = 2, \ldots, m - 2. \tag{2.23}$$

On the other hand, from the equation (II.2.18) of Codazzi, we have

$$(\nabla_X A_x)Y - \omega_x^y(X)A_y(Y) = (\nabla_Y A_x)X - \omega_x^y(Y)A_y(X). \tag{2.24}$$

From (2.23) and (2.24), we obtain

$$\omega_x^1(X)A_1(Y) - \omega_x^1(Y)A_1(X) = 0, \quad \text{for } x = 2, \ldots, m - 2. \tag{2.25}$$

Thus, by the fact that trace $A_1 = 0$ and $A_1 \neq 0$ on M_1, we have

$$\omega_x^1 = -\omega_1^x = 0. \tag{2.26}$$

From (2.23) and (2.26) we see that every component of M_1 is contained in a great 3-sphere of $R^m(c)$.

Case (a) $\text{Int}(M_0) = \emptyset$. If M_1 contains only one component, then $M = \partial M_1 \cup M_1$ is contained in a great 3-sphere of $R^3(c)$. In this case, there is nothing to prove. If M_1 contains more than one component, then $M_0 = \partial M_1$ and M_0 separates M_1 into its different components. Let U_1 and U_2 be two components of M_1 such that $\partial U_1 \cap \partial U_2$ is nonempty. Then, by the fact that $K^N = 0$ and Proposition 1.1, we see that, for any $P \in \partial U_1 \cap \partial U_2$, there exists a neighborhood W of P in M and a field of orthonormal normal frames η_x on W such that the η_x are parallel. Let ξ_x be a field of orthonormal normal frames in M_1 with $\xi_1 \in N'$. Since our discussion is only on a small neighborhood of P, we may assume that the ξ_x are defined on the whole of $W \cap M_1$. Because ξ_1 is parallel in the normal bundle and the normal connection is flat, we may assume that the ξ_x are all parallel on $W \cap M_1$. Since the ξ_x and η_x are all parallel, if we put

$$\eta_x = a_x^y \xi_y, \tag{2.27}$$

then a_x^y are constants on each component of $W \cap M_1$. Now, let c_x^y be the constants defined by $c_x^y = a_x^y|_{U_1 \cap W}$. We put

$$\bar{\eta}_x = c_x^y \eta_y,$$

on W. Then $\bar{\eta}_x$ is a field of orthonormal normal frames on W such that the $\bar{\eta}_x$ are parallel and $\bar{\eta}_x = \xi_x$ on $U_1 \cap W$. since $A_{\xi_x} = 0$, for $x = 2, \ldots, m-2$, we have

$$A_{\bar{\eta}_x} = 0 \qquad \text{on } (U_1 \cap W) \cup (M_0 \cap W),$$
$$A_{\bar{\eta}_x} = c_x^y a_y^1 A_{\xi_1}, \quad \text{on } U_2 \cap W, \tag{2.28}$$

for $x = 2, \ldots, m-2$. On the other hand, since the $\bar{\eta}_x$ are parallel and $M_1(\bar{\eta}_x) = 0$, from Lemma 2.2, the functions

$$\varphi(\bar{\eta}_x) = \frac{1}{2} \tilde{g}(L - N, \bar{\eta}_x) - \tilde{g}(M, \bar{\eta}_x) i$$

are analytic. Hence, by (2.28), we see that

$$c_x^y a_y^1 \tilde{g}(L - N, \xi_1) - 2 c_x^y a_y^1 \tilde{g}(M, \xi_1) i = 0, \quad x = 2, \ldots, m-2, \tag{2.29}$$

on $U_2 \cap W$. Since $\tilde{g}(L - N, \xi_1) \neq 0$ on $U_2 \cap W$, we have

$$c_x^y a_y^1 = 0, \quad x = 2, \ldots, m-2 \tag{2.30}$$

on $U_2 \cap W$. This implies that $A_{\bar{\eta}_2} = \cdots = A_{\bar{\eta}_{m-2}} = 0$ on $U_2 \cap W$. Hence, we have $\bar{\eta}_1 = \xi_1$ on $U_2 \cap W$. From this fact we see that U_1 and U_2 are both contained in the same great 3-sphere of $R^m(c)$. Consequently, we see that the surface M is contained in a great 3-sphere of $R^m(c)$.

Case (b) Int$(M_0) \neq \emptyset$. In this case, each component of Int(M_0) is contained in a great 2-sphere of $R^m(c)$. By an argument analogous to the proof of case (a), we may claim that M is contained in a great 3-sphere of $R^m(c)$.

Case (iii) $H \neq 0$. Since the mean curvature vector H is nonzero and it is parallel in the normal bundle, we may choose the ξ_x in such a way that

$$\xi_1 = H/|H| \tag{2.31}$$

and

$$\xi_2 \in N' \text{ on } M_1 = \{P \in M; \dim N' = 1 \text{ at } P\}. \tag{2.32}$$

Then we have

$$A_3 = \cdots = A_{m-2} = 0, \quad \omega_1^x = \omega_x^1 = 0,$$
$$A_x = A_{\xi_x}. \tag{2.33}$$

From (2.33) and the equation (2.24) of Codazzi we have

$$\omega_x^2(X)A_2(Y) - \omega_x^2(Y)A_2(X) = 0, \quad \text{on } M_1, \tag{2.34}$$

for $x = 3, \ldots, m - 2$. Thus, by the fact that $A_2 \neq 0$ and trace $A_2 = 0$ on M_1, we find

$$\omega_x^2 = 0, \quad \text{on } M_1. \tag{2.35}$$

Hence, from (2.33) and (2.35), we see that every component of M_1 is contained in a great 4-sphere of $R^m(c)$.

If $\text{Int}(M_0) \neq \emptyset$, then, by the fact that $\omega_x^1 = 0$ and dim $N' = 0$, we see that every component of the closure of $\text{Int}(M_0)$ is contained in a great 3-sphere of $R^m(c)$. If we have either $M_0 = \emptyset$ or $M_1 = \emptyset$, then from the above observation, we know that the surface M must be contained in a great 4-sphere of $R^m(c)$. If both M_0 and M_1 are nonempty, then it is clear that M will be contained in a great 4-sphere of $R^m(c)$, if, for every point $P \in \partial M_0$, there exists a neighborhood of the point P which is contained in a great 4-sphere of $R^m(c)$.

For every point $P \in \partial M_0$, P must belong either to some $\partial U_1 \cap \partial V$ or to some $\partial U_1 \cap \partial U_2$, where U_1, U_2 are components of M_1 and V is a component of $\text{Int}(M_0)$. Therefore, we may consider these two cases separately.

Case (c) $P \in \partial U_1 \cap \partial V$. Since $P \in \partial V$ and the closure of V is contained in a great 3-sphere of $R^m(c)$, the closure of V is contained in the great 3-sphere corresponding to the 3-dimensional subspace of $T_P(R^m(c))$ spanned by the mean curvature vector H_P at the point P and the tangent space $T_P(M)$ of M at P (under the exponential map $\exp_P$). Therefore, by the fact that $P \in \partial U_1$, we see that this great 3-sphere of $R^m(c)$ is contained in the great 4-sphere of $R^m(c)$ which contains ∂U_1.

Case (d) $P \in \partial U_1 \cap \partial U_2$. Since the normal connection of M in $R^m(c)$ is flat, there exists a neighborhood W of the point P such that there exist $m - 2$ mutually orthogonal unit normal vector fields η_x defined on W and each of the η_x is parallel in the normal bundle. Without loss of generality we may assume that the ξ_x are also parallel in the normal bundle and that they are defined on the whole of $W \cap M_1$. In exactly the same way as in the proof of case (a), we may assume that $\eta_x = \xi_x$ on the set $U_1 \cap W$. Since ξ_1 and η_1 are both parallel on W and they are equal on $U_1 \cap W$, $\eta_1 = \xi_1$ everywhere on W. Hence, we obtain $\eta_1 = H/|H|$ on the whole of W. Thus, we may put

$$\eta_x = \sum_{y=2}^{m-2} a_x^y \xi_y \quad \text{for } x = 2, \ldots, m - 2, \tag{2.36}$$

on W. Since $A_3 = \cdots = A_{m-2} = 0$ on M_1, we have

$$\begin{aligned} A_{\eta_x} &= 0 && \text{on } (U_1 \cap W) \cup (M_0 \cap W), \\ A_{\eta_x} &= a_x^2 A_2 && \text{on } U_2 \cap W, \end{aligned} \qquad (2.37)$$

for $x = 3, \ldots, m - 2$. Thus, by Lemma 2.2, we see that

$$\varphi(\eta_x) = \frac{1}{2}\tilde{g}(L - N, \eta_x) - \tilde{g}(M, \eta_x)i = 0 \quad \text{for } x = 3, \ldots, m - 2, \qquad (2.38)$$

on W. Hence, we obtain

$$\frac{1}{2}a_x^2 \tilde{g}(L - N, \eta_x) - a_x^2 \tilde{g}(M, \eta_x)i = 0 \quad \text{for } x = 3, \ldots, m - 2,$$

on $U_2 \cap W$. Since trace $A_2 = 0$ and $A_2 \neq 0$ on M_1, we obtain from this equation

$$a_x^2 = 0 \quad \text{for } x = 3, \ldots, m - 2.$$

This implies that

$$\eta_1 = \xi_1 \quad \text{and} \quad \eta_2 = \xi_2 \text{ or } -\xi_2$$

on the whole of $W \cap M_1$. From this and (2.33) and (2.35) we see that the closure of $U_1 \cup U_2$ is contained in a great 4-sphere of $R^m(c)$, that is, the closure of U_1 and the closure of U_2 are contained in the same great 4-sphere of $R^m(c)$. This completes the proof of the lemma. □

Now, *we return to the proof of the theorem.* Since the mean curvature vector H of M is parallel in the normal bundle, H has constant length. If H vanishes everywhere, then M is a minimal surface of $R^m(c)$. If H is nonzero and M is not a minimal surface of any small hypersphere of $R^m(c)$, then, by Lemma 2.4, we see that the normal connection of M in $R^m(c)$ is flat. Hence, by applying Lemmas 2.3 and 2.5, we see that M is contained in a 3-sphere of $R^m(c)$ with constant mean curvature. This completes the proof of the theorem. □

Remark 2.1. From the proof of Lemma 2.5, we see that if M is a minimal surface of an m-dimensional space form $R^m(c)$ with flat normal connection, then M is contained in a great 3-sphere of $R^m(c)$.

As corollaries of Lemma 2.2 we have the following

Proposition 2.1 (Ferus, 1971b; Hoffman, 1972, 1973; Ruh, 1971; Smyth, 1973). *Let M be a closed oriented surface of genus zero in an m-dimensional space form $R^m(c)$. If there exists a parallel isoperimetric section on M, then M is contained in a hypersphere of $R^m(c)$.*

Proposition 2.2 (Ferus, 1971; Hoffman, 1972, 1973; Ruh, 1971; Smyth, 1973). *Let M be a closed oriented surface of genus zero in an m-dimensional space form $R^m(c)$ with nonzero parallel mean curvature vector. Then M is contained in a* *small hypersphere of positive curvature in $R^m(c)$ as a minimal surface. When the codimension is two, M is contained in a small 2-sphere of $R^m(c)$.*

These two propositions follow immediately from Lemma 2.2, Problem II.7, Proposition II.3.2, Proposition II.5.5, and the fact that every analytic quadratic differential on a closed oriented surface of genus zero is identically zero.*

Remark 2.2. If the codimension is one, Propositions 2.1 and 2.2 were obtained, respectively, by Hopf (1951) and Almgren (1966).

3 Surfaces with Constant Mean Curvature in $R^3(c)$

In this section we shall assume that the ambient space N is a 3-dimensional space form $R^3(c)$ of curvature c.

Proposition 3.1. *Let M be a surface in a 3-dimensional space form $R^3(c)$ with constant mean curvature $|H|$. If M has nonzero constant Gaussian curvature G, then M is contained in a hypersphere of $R^3(c)$.*

Proof. Since the codimensions is one and M has constant mean curvature, the mean curvature vector H of M is parallel in the normal bundle. If $H = 0$, then, by Problem III.4, we see that M is contained in a great hypersphere of $R^3(c)$. If $H \neq 0$, and $H = |H|\xi_1$, then, by Lemma 2.2, we see that the function

$$\varphi = \frac{1}{2}g(L - N, \xi_1) - g(M, \xi_1)i \tag{3.1}$$

is analytic on $z = x^1 + ix^2$ for every set of isothermal coordinates x^1, x^2. From (3.1) and the equation of Gauss, we find

$$|\varphi|^2 = \varphi\bar{\varphi} = E^2\{|H|^2 - G + c\}, \tag{3.2}$$

where G is the Gaussian curvature of M and E is given by $g = E\{(dx^1)^2 + (dx^2)^2\}$.

Suppose that $G = |H|^2 + c$. Then φ is identically zero. Hence, M is totally umbilical in $R^3(c)$. By Proposition II.3.2, we see that M is an open piece of a small hypersphere of $R^3(c)$.

Now, suppose that $G \not\equiv |H|^2 + c$. Then, by (3.2), $\varphi \not\equiv 0$. Since φ is analytic, $\log|\varphi|^2$ is harmonic, that is,

$$\Delta \log|\varphi|^2 = 0. \tag{3.3}$$

On the other hand, by Problem I.14, we see that, if $\varphi \neq 0$, the Gaussian curvature G of M is given by

$$-G = \frac{1}{4E}\Delta \log E^2 = \frac{1}{4E}\Delta \log\left(\frac{1}{|H|^2 - G + c}\right). \tag{3.4}$$

---

*This fact follows from the famous Riemann-Roch's theorem.

Thus, by the constancy of $|H|$ and G and Eq. (3.3), we find that the Gaussian curvature G of M vanishes. This is a contradiction. Thus, we have proved the proposition completely.   □

Combining Problem II.8, Theorem 2.1, and Proposition 3.1, we obtain the following

Proposition 3.2 (Chen and Ludden, 1972a,b; Hoffman, 1972, 1973). *The minimal surfaces of a small hypersphere of a euclidean m-space E^m, the open pieces of the product of two plane circles and the open pieces of a circular cylinder are the only nonminimal surfaces in E^m with parallel mean curvature vector and constant Gaussian curvature.*

Proposition 3.3 (Klotz and Osserman, 1966, 1967; Hoffman, 1972, 1973). *Let M be a complete surface in a 3-dimensional space form $R^3(c)$ $(c = 0, 1)$ with constant mean curvature $H \neq 0$. If the Gaussian curvature G of M does not change sign, then either M is a small hypersphere of $R^3(c)$ or M is flat.*

Proof. Let x^1, x^2 be isothermal coordinates in M. Then, by Lemma 2.2, the function

$$\varphi = \frac{1}{2}\tilde{g}(L - N, \xi_1) - \tilde{g}(M, \xi_1)i \tag{3.5}$$

is analytic in $z = x^1 + x^2 i$. We now consider the cases $G \leq 0$ and $G \geq 0$ separately.

Case (i) $G \leq 0$. From (3.2), we have

$$|\varphi|^2 = E^2(|H|^2 - G + c) > 0. \tag{3.6}$$

Since $\varphi \neq 0$ is analytic, $\log |\varphi|^2$ is harmonic, that is, (3.3) holds. Let g^* denote the new metric tensor in M given by

$$g^* = |\varphi|\{(dx^1)^2 + (dx^2)^2\}.$$

Then, by (3.3), we see that the Gaussian curvature G^* of M with respect to the metric g^* vanishes. Thus, g^* is a flat metric in M which is conformally equivalent to g. Thus, the universal covering surface $\tilde{M}$ of M is conformally equivalent to the euclidean plane E^2. On the surface M, the function

$$\log \frac{|\varphi|}{E} = \log |\varphi| - \log E$$

is a globally defined function and it is bounded from below by $\log |H| \in \mathcal{R}$ by virtue of (3.6). Moreover, from (3.3), we find

$$\Delta \log \frac{|\varphi|}{E} = -\Delta \log E = 2EG \leq 0, \tag{3.7}$$

that is, the function $\log(|\varphi|/E)$ is superharmonic. Lifting this function to the universal covering surface $\tilde{M}$, we obtain a superharmonic function bounded from below on $\tilde{M}$. Since $\tilde{M}$ is conformally equivalent to the euclidean plane, $\tilde{M}$ is parabolic. Hence, we see that $\log(|\varphi|/E)$ is a constant. This implies that the Gaussian curvature G of M is zero.

Case (ii) $G \geq 0$. By a theorem of Huber (1957), we know that a complete surface with $G \geq 0$ is either compact or parabolic. Suppose that M is compact. If $G \neq 0$, then, by the well-known theorem of Gauss-Bonnet, we know that M is of genus zero. Hence, by Proposition 2.2, we see that M is a small hypersphere of $R^3(c)$. Now, suppose that M is parabolic and noncompact. As in case (i) we see that (3.3) holds. Hence, we have

$$0 = \Delta \log |\varphi|^2 = 2 \left(\Delta \log E + \Delta \log \frac{|\varphi|}{E} \right)$$
$$= 2 \left(-2GE + \Delta \log \frac{|\varphi|}{E} \right)$$
$$\leq 2\Delta \log \frac{|\varphi|}{E}. \tag{3.8}$$

This implies that $\log(|\varphi|/E)$ is a subharmonic function. On the other hand, we have

$$\log \frac{|\varphi|}{E} = \frac{1}{2} \log(|H|^2 - G + c)$$
$$\leq \frac{1}{2} \log(|H|^2 + c).$$

Therefore, $\log(|\varphi|/E)$ is bounded from above. Consequently, we see that $(|\varphi|/E)$ is constant on M since M is parabolic. This implies that the Gaussian curvature G is also constant. Applying Proposition 3.1, we obtain the proposition. □

Combining Theorem 2.1, Proposition 3.2, and Proposition 3.3, we obtain the following

Proposition 3.4 (Chen, 1973c; Hoffman, 1972, 1973). *Let M be a complete surface in a euclidean m-space E^m with parallel mean curvature vector $H \neq 0$. If the Gaussian curvature of M does not change sign, then M is one of the following surfaces:*

(i) *a minimal surface of a small hypersphere of E^m,*

(ii) *a product surface of two plane circles, or*

(iii) *a product surface of a straight line and a plane circle, that is, a circular cylinder.*

Remark 3.1. Closed hypersurfaces of a euclidean $(n+1)$-space E^{n+1} of constant mean curvature satisfy a variational principle. Namely, a closed hypersurface M of E^m has constant mean curvature if and only if its n-dimensional volume is stable with respect to $(n+1)$-dimensional volume preserving variations, where the $(n+1)$-dimensional volume is the volume in E^{n+1} enclosed by M (Hopf).

4   Local Existence Theorem for Surfaces with Constant Mean Curvature

From the discussion of §2, we know that if M is a surface in a 3-dimensional space form $R^3(k)$ of curvature k with constant mean curvature $\alpha = |H|$ and isothermal coordinates $\{x^1, x^2\}$, then there exists an analytic function φ given by (2.11). In this section, we shall construct a surface in a 3-dimensional space form $R^3(k)$ with constant curvature by using a positive constant α and an analytic function φ of $z = x + iy$ defined in a neighborhood of the origin in the (x, y)-plane. This construction was obtained by Wolf (1967) (see also Hoffman, 1972, 1973; Chen, 1973c) for the euclidean case.

Theorem 4.1. *Let $\varphi \neq 0$ be an analytic function of $z = x + iy$ defined in a neighborhood of the origin in the (x, y)-plane. Let α be a positive number. Then there exists a neighborhood Ω of the origin, an isothermal metric $E(dx^2 + dy^2)$ defined on Ω, and an isometric immersion of (Ω, E) into a 3-dimensional space form $R^3(k)$ of curvature k such that the immersed surface has constant mean curvature α and φ is the function given by (2.11).*

Proof. We consider the partial differential equation

$$\frac{\partial^2 f}{\partial x^2} + \frac{\partial^2 f}{\partial y^2} = 2\{|\varphi|^2 e^{-f} - (\alpha^2 + k)e^f\} \tag{4.1}$$

with the initial conditions

$$f(0, y) = 0, \quad \frac{\partial f}{\partial x}(0, y) = 0. \tag{4.2}$$

By a theorem of Cauchy-Kovalewski [see, for instance, Courant and Hilbert (1962, vol. 2, p. 39)] we see that the solution f of this partial differential equation exists on a neighborhood Ω of the origin $(0,0)$ and is unique. Put $E = e^f$. Then (4.1) gives

$$E^2(\alpha^2 + k - G) = |\varphi|^2, \tag{4.3}$$

where $G = -(1/2E)\Delta \log E$. We consider the isothermal coordinates x, y with the metric tensor $g = E(dx^2 + dy^2)$ on Ω. Put

$$N = \Omega \times \mathcal{R},$$

where $\mathcal{R}$ is the real line. Then N can be considered as the total space of a vector bundle over Ω with the real line as the fibres. With the usual product on $\mathcal{R}$, N becomes a Riemannian line bundle with a usual connection D. Let ξ be the natural unit vector field of N. We define $X_1 = \partial/\partial x$, $X_2 = \partial/\partial y$, and a bilinear mapping h by

$$h(X_1, X_1) = (\alpha + \operatorname{Re} \varphi)\xi,$$
$$h(X_1, X_2) = h(X_2, X_1) = -(\operatorname{Im} \varphi)\xi,$$
$$h(X_2, X_2) = (\alpha - \operatorname{Re} \varphi)\xi.$$

Then, by the Existence Theorem, we see that there exists an isometric immersion of Ω with the metric tensor g into a 3-dimensional space form $R^3(k)$ of curvature k with N as the normal bundle of Ω in $R^3(k)$ and h as the second fundamental form if and only if g and h satisfy the equation (II.2.15) of Gauss, the equation (II.2.18) of Codazzi, and the equation (II.2.16) of Ricci. Since N is a line bundle, the equation (II.2.16) of Ricci is automatically satisfied. The equation (II.2.15) of Gauss follows immediately from Eq. (4.3), and the equation (II.2.18) of Codazzi reduces to the Cauchy-Riemann equations of the analytic function φ. Consequently, we obtain the theorem. $\qquad\square$

Remark 4.1. By the Rigidity Theorem, we see that the isometric immersion given in Theorem 4.1 is unique up to rigid motions in $R^3(k)$, that is, up to the isometries of $R^3(k)$.

5   Surfaces with Parallel Minimal Section

Let M be an n-dimensional submanifold of a Riemannian manifold N. For a unit normal vector field ξ of M in N, if we have $M_1(\xi) = 0$ identically, then ξ is called a *minimal section* on M; if A_ξ is not proportional to the identity transformation everywhere, then ξ is called a *umbilical-free section* on M; if the determinant of A_ξ is nowhere zero, then ξ is called a *nondegenerate* section; and if A_ξ is not identically zero, then ξ is called a *nongeodesic section*.

Proposition 5.1. *Let M be a closed surface in an m-dimensional space form $R^m(c)$ such that the Gaussian curvature G of M does not change its sign. If there exists a parallel umbilical-free isoperimetric section on M, then M is flat and $M_2(\xi)$ is constant.*

Proof. By the assumption, ξ is a parallel umbilical-free isoperimetric section, we have $M_1(\xi)^2 - M_2(\xi) \neq 0$ everywhere and, by Lemma 2.2, the function

$$\varphi = \left\langle \frac{L-N}{2}, \xi \right\rangle - \langle M, \xi \rangle i \tag{5.1}$$

is analytic in $z = x^1 + ix^2$ for every set of isothermal coordinates x^1, x^2. Thus, the function $\log |\varphi|^2$ is harmonic, that is,

$$\Delta \log |\varphi|^2 = 0. \tag{5.2}$$

Since

$$|\varphi|^2 = \varphi\bar\varphi = E^2\{M_1(\xi)^2 - M_2(\xi)\},$$

we have

$$\Delta \log E^2 = -\Delta \log\{M_1(\xi)^2 - M_2(\xi)\}. \tag{5.3}$$

By Problem I.14, the Gaussian curvature G of M is given by

$$G = -\frac{1}{4E}\Delta \log E^2. \tag{5.4}$$

Hence, by (5.3) and (5.4), we find that

$$G = \frac{1}{4E}\Delta \log\{M_1(\xi)^2 - M_2(\xi)\}. \tag{5.5}$$

By the assumption, G does not change its sign. Hence, (5.5) implies that the function $\log\{M_1(\xi)^2 - M_2(\xi)\}$ is either a subharmonic function or a superharmonic function on M. Thus, by Hopf's lemma, we see that $\{M_1(\xi)^2 - M_2(\xi)\}$ is a constant. From this we see that the Gaussian curvature G of M vanishes and $M_2(\xi)$ is constant. This proves the proposition. □

Proposition 5.2. *Let M be a closed surface of a 4-dimensional space form $R^4(c)$ of curvature $c \leq 0$ such that the Gaussian curvature G of M does not change its sign. Suppose that there exists a parallel nondegenerate minimal section on M. Then the mean curvature vector H of M is parallel in the normal bundle.*

Proof. Without loss of generality, we may assume that ξ_1, ξ_2 are two mutually orthogonal unit normal vector fields of M in $R^4(c)$ such that ξ_2 is the parallel nondegenerate minimal section. Then we have $\nabla^\perp_X \xi_x = 0$ for any vector field X in M. By the equation (II.2.7) of Codazzi, we have

$$(\nabla_X h^x)(Y, Z) = (\nabla_Y h^x)(X, Z) \tag{5.6}$$

for any vector fields X, Y, Z in M. By Proposition 1.2, we may assume that E_1, E_2 are the principal directions of ξ_1 and ξ_2 with the principal curvatures λ_1, λ_2 and μ_1, μ_2, respectively. Since ξ_2 a minimal section, $\mu_1 + \mu_2 = 0$. By Proposition 5.1, we have $\mu_1\mu_2 = $ constant and $\lambda_1\lambda_2 = -\mu_1\mu_2 - c > 0$. Consequently, we see that the second fundamental tensors of M with respect to ξ_1 and ξ_2 are given, respectively, by

$$A_1 = \begin{pmatrix} \lambda_1 & 0 \\ 0 & \lambda_1 \end{pmatrix}, \quad A_2 = \begin{pmatrix} \mu & 0 \\ 0 & -\mu \end{pmatrix}, \quad \mu \neq 0, \tag{5.7}$$

where $\lambda_1\lambda_2 = \mu^2 - c$ is a positive constant.

From (5.6) and (5.7) we find

$$h^2(\nabla_{E_k} E_i, E_j) + h^2(E_i, \nabla_{E_k} E_j) = h^2(\nabla_{E_i} E_k, E_j) + h^2(E_k, \nabla_{E_i} E_j).$$

Hence, if we put $\nabla_{E_i} E_j = \omega_j^l(E_i)E_l$, we have

$$\{\omega_i^j(E_k) - \omega_k^j(E_i)\}h^2(E_j, E_j) = \omega_j^k(E_i)h^2(E_k, E_k) - \omega_j^i(E_k)h^2(E_i, E_i). \tag{5.8}$$

Setting $j = i$ and $k \neq i$ in (5.8), we find

$$\omega_k^i(E_i) = 0 \quad \text{for } k \neq i \ (i \text{ not summed}) \tag{5.9}$$

by virtue of (5.7) and the skew symmetry of ω_i^j. Similarly, by setting $k = j$ and $j \neq i$ in (5.8), we find

$$\omega_i^k(E_k) = -\omega_k^i(E_k) = 0 \quad \text{for } k \neq i \ (k \text{ not summed}). \tag{5.10}$$

Combining (5.9) and (5.10) and using $\omega_i^i = 0$ for $i = 1, 2$, we find

$$\nabla_{E_i} E_j = 0. \tag{5.11}$$

Setting $x = 1$ in (5.6) and applying (5.11), we find

$$\nabla_{E_k} h^1(E_i, E_j) = \nabla_{E_i} h^1(E_k, E_j). \tag{5.12}$$

Setting $i = j$ and $k \neq i$ in (5.12) and applying (5.7), we have

$$\nabla_{E_k} h^1(E_i, E_i) = 0 \quad \text{for } k \neq i. \tag{5.13}$$

On the other hand, since

$$M_2(\xi_1) = h^1(E_i, E_i)h^1(E_k, E_k)$$

is a nonzero constant for $i \neq k$, by applying (5.13), we find

$$\begin{aligned}
0 &= \nabla_{E_k}\{h^1(E_i, E_i)h^1(E_k, E_k)\} \\
&= h^1(E_i, E_i)\nabla_{E_k} h^1(E_k, E_k)
\end{aligned} \tag{5.14}$$

for $k \neq i$. From this we find

$$\nabla_{E_k} h^1(E_k, E_k) = 0. \tag{5.15}$$

Combining this with (5.13), we see that the principal curvatures λ_1 and λ_2 of the normal direction ξ_1 are constants. Hence, in particular, $M_1(\xi_1) = (\lambda_1 + \lambda_2)/2$ is a constant. From this, we see that the mean curvature vector H of M has constant length and hence, by parallelism of ξ_1 in the normal bundle, we see that the mean curvature vector H is also parallel in the normal bundle. This proves the proposition. $\qquad\square$

Combining Proposition 3.2, Proposition 5.1, and Proposition 5.2, we obtain the following characterization result for the product surface of two plane circles.

Proposition 5.3. *Let M be a closed surface in a euclidean 4-space E^4 such that the Gaussian curvature of M does not change its sign. Then M is the product surface of two plane circles if and only if there exists a parallel nondegenerate minimal section on M.*

If the Gaussian curvature of M vanishes, then we have the following local result:

Proposition 5.4. *Let M be a surface in a euclidean 4-space E^4 with zero Gaussian curvature. Then M is an open piece of the product surface of two plane circles if and only if there exists a parallel minimal nongeodesic section on M.*

The proof of this proposition is similar to that of Proposition 5.3 and hence omitted.

Remark 5.1. If $c = 0$, Proposition 5.1 was obtained independently by Chen (1973a), Houh (1972c) and Wegner (1974a).

6 Axiom of Spheres

In his book on Riemannian geometry, Élie Cartan (1946) defined the axiom of n-planes as follows. A Riemannian manifold N of dimension $m \geq 3$ satisfies the *axiom of n-planes* if, for each point $P \in N$ and any n-dimensional subspace T'_P of the tangent space $T_P(N)$, there exists an n-dimensional totally geodesic submanifold M containing P such that the tangent space of M at P is T'_P, where n is a fixed integer $2 \leq n < m$. He proved that if M satisfies the axiom of n-planes for some n, then N is a space of constant curvature (Cartan, 1946).

Recently, Leung and Nomizu generalized Cartan's idea and proposed the

Axiom of n-spheres. For each point $P \in N$ and any n-dimensional subspace T'_P of $T_P(N)$, there exists an n-dimensional totally umbilical submanifold M with parallel mean curvature vector such that $P \in M$ and $T_P(M) = T'_P$.

They proved that

Theorem 6.1 (Leung and Nomizu, 1971). *If a Riemannian manifold N of dimension $m \geq 3$ satisfies the axiom of n-spheres for some n, $2 \leq n < m$, then N is a space of constant curvature.*

Proof. We first prove the following lemmas.

Lemma 6.1 (Cartan, 1946). *Let N be an m-dimensional Riemannian manifold $(m > 2)$. If we have*

$$\tilde{K}(\tilde{X}, \tilde{Y}; \tilde{Z}, \tilde{X}) = 0$$

for any orthonormal vector fields $\tilde{X}$, $\tilde{Y}$, $\tilde{Z}$ in N, then N has constant sectional curvature.

Proof of Lemma 6.1. Let $\tilde{Y}' = (\tilde{Y} + \tilde{Z})/\sqrt{2}$ and $\tilde{Z}' = (\tilde{Y} - \tilde{Z})/\sqrt{2}$. Since $\tilde{X}$, $\tilde{Y}'$, $\tilde{Z}'$ are again orthonormal, we have

$$\tilde{K}(\tilde{X}, \tilde{Y}'; \tilde{Z}', \tilde{X}) = 0,$$

from which we find

$$\tilde{K}(\tilde{X}, \tilde{Y}; \tilde{Y}, \tilde{X}) = \tilde{K}(\tilde{X}, \tilde{Z}; \tilde{Z}, \tilde{X}).$$

This means that the sectional curvature for the plane section $\gamma(\tilde{X}, \tilde{Y})$ is equal to that of the plane section $\gamma(\tilde{X}, \tilde{Z})$. From this we see that N has constant sectional curvature. □

Lemma 6.2 (Equation of Codazzi). *Let M be an n-dimensional submanifold of an m-dimensional Riemannian manifold. Then for any vector fields X, Y in M and any normal vector field ξ, the tangential component of $\tilde{K}(X, Y)\xi$ is equal to*

$$(\nabla_Y A_\xi)(X) - (\nabla_X A_\xi)(Y) + A_{\nabla^\perp_X \xi}(Y) - A_{\nabla^\perp_Y \xi}(X).$$

This lemma follows immediately from the equation (II.2.7) of Codazzi.

Now, we *return to the proof of the theorem.* Let P be any point in N and X, Y, Z be three orthonormal vectors in $T_P(N)$. Let T'_P be an n-dimensional subspace of $T_P(N)$ containing X and Y and normal to Z. By the axiom of n-spheres, there exists an n-dimensional totally umbilical submanifold M with parallel mean curvature vector H such that $P \in M$ and $T_P(M) = T'_P$. Let U be a normal neighborhood with a normal coordinate system around P. For each $Q \in U$, let η_Q be the normal vector at Q to M which is parallel to Z with respect to the normal connection $\nabla^\perp$ along the geodesic from P to Q in U. Along each geodesic we have $\tilde{g}(H, \eta) = $ constant, say λ, so that $A_\eta = \lambda I$ at every point of U. Thus, we have

$$\nabla_X A_\eta = \nabla_Y A_\eta = 0 \text{ at } P.$$

We have also

$$\nabla^\perp_X \eta = \nabla^\perp_Y \eta = 0 \text{ at } P.$$

Thus, from Lemma 6.2, we see that the tangential component of $\tilde{K}(X, Y)Z$ is zero. In particular, we have $\tilde{K}(X, Y; Z, X) = 0$. Thus, by applying Lemma 6.1, we see that N has constant sectional curvature. This proves the theorem. □

7 Submanifolds with Nonnegative Sectional Curvatures

In this section we shall assume that M is an n-dimensional submanifold of an m-dimensional space form $R^m(c)$ of curvature c with parallel mean curvature vector H.

Let ξ_x be orthonormal normal vector fields of M with ξ_1 in the direction of the mean curvature vector. Then we have

$$h^{(y)} = 0 \quad \text{for } y = 2, \ldots, m - n, \tag{7.1}$$

where $h^{(y)} = (1/n)h_t^{ty}$. From the definition of $\overline{\nabla}h$, we have

$$\overline{\nabla}_X h^{(1)} = X h^{(1)}, \quad \overline{\nabla}_X h^{(y)} = h^{(1)}\omega_{n+1}^{n+y}(X), \quad y = 2, \ldots, m - n, \tag{7.2}$$

where the ω_{x+n}^{y+n} are given by $\nabla_X^{\perp}\xi_x = \omega_{x+n}^{y+n}(X)\xi_y$.

By the parallelism of the mean curvature vector H in the normal bundle, we find from (7.2) that

$$\overline{\nabla}_X h^{(1)} = 0, \quad \overline{\nabla}_X h^{(y)} = 0 \quad \text{for } y = 2, \ldots, m - n. \tag{7.3}$$

Substituting (7.3) into (II.6.3), we find

$$\overline{\nabla}_Y \overline{\nabla}_X h^{(x)} = 0 \quad \text{for } x = 1, 2, \ldots, m - n. \tag{7.4}$$

Now, suppose that the normal connection $\nabla^{\perp}$ is flat, then, by Proposition 1.2, we have

$$[A_x, A_y] = 0. \tag{7.5}$$

Hence, by substituting (7.4) and (7.5) into (III.5.4), we find

$$\frac{1}{2}\Delta\langle h \rangle^2 = nc\langle h \rangle^2 - n^2 c|H|^2 - \sum_{x,y}(\text{trace}(A_x A_y))^2 + \sum_x (nh^{(1)})\text{trace}(A_x A_1 A_x)$$

$$+ (\overline{\nabla}_k h_{ji}^x)(\overline{\nabla}^k h_x^{ji}), \tag{7.6}$$

where $\langle h \rangle$ is the length of the second fundamental form and Δ is the Laplacian of M.

On the other hand, by (7.5), we know that all of the second fundamental tensors A_x are simultaneously diagonalizable. Hence, we may choose orthonormal vector fields E_i in M such that the E_i are the principal directions with respect to each of the ξ_x with principal curvatures ρ_i^x, respectively. From (7.6), we have

$$\frac{1}{2}\Delta\langle h \rangle^2 = nc\sum_{x,i}(\rho_i^x)^2 - c\left(\sum_i \rho_i^1\right)^2 - \sum_{x,y}\left(\sum_i \rho_i^x \rho_i^y\right)^2$$

$$+ \left(\sum_x \rho_i^1\right)\left(\sum_{x,i} \rho_i^1(\rho_i^x)^2\right) + (\overline{\nabla}_k h_{ji}^x)(\overline{\nabla}^k h_x^{ji})$$

$$= \sum_{i,j,x,y}\{\rho_i^y \rho_j^y(\rho_j^x)^2 - \rho_i^y \rho_j^y \rho_i^x \rho_j^x\} + c\sum_{x,i,j} \rho_i^x(\rho_i^x - \rho_j^x) + |\overline{\nabla}h|^2$$

$$= \sum_{i,j} \sum_y \{\rho_i^y \rho_j^y\} \sum_x \rho_j^x (\rho_j^x - \rho_i^x) + c \sum_{i \leqq j} \sum_y \rho_i^x (\rho_i^x - \rho_j^x) + |\overline{\nabla} h|^2$$

$$= \sum_{i \leqq j} \left(\sum_y \rho_i^y \rho_j^y \right) \sum_x (\rho_i^x - \rho_j^x)^2 + c \sum_{i \leqq j} \sum_x \rho_i^x (\rho_i^x - \rho_j^x) + |\overline{\nabla} h|^2$$

$$= \sum_{i \leqq j} \left\{ \sum_x (\rho_i^x - \rho_j^x)^2 \left(c + \sum_y \rho_i^y \rho_j^y \right) \right\} + |\overline{\nabla} h|^2, \tag{7.7}$$

where $|\overline{\nabla} h|^2 = (\overline{\nabla}_k h_{ji}^x)(\overline{\nabla}^k h_x^{ji})$.

Lemma 7.1 (Erbacher, 1972; Yano and Ishihara, 1971). *Suppose that M has nonnegative sectional curvature, H is parallel, and the normal connection is flat. If M has constant scalar curvature or M is compact, then $\Delta \langle h \rangle^2 = 0$ and $|\overline{\nabla} h| = 0$. If M is compact, M has constant scalar curvature.*

Proof. If M has constant scalar curvature, then, by the constancy of the mean curvature, we see that h has constant length $\langle h \rangle$. Hence, $\Delta \langle h \rangle^2 = 0$. On the other hand, since the sectional curvature of the plane section $\gamma(E_i, E_j)$ is given by $c + \sum_y \rho_i^y \rho_j^y$, by (7.7) we have $|\overline{\nabla} h| \leqq 0$. Thus we have $|\overline{\nabla} h| = 0$. If M is compact, by (7.7) and Green's theorem, we also have $\Delta \langle h \rangle^2 = 0$. Hence, $|\overline{\nabla} h| = 0$, and by the constancy of mean curvature we see that the scalar curvature is also constant. This proves the lemma. $\square$

Lemma 7.2 (Erbacher, 1972, Yano and Ishihara, 1971). *If the assumptions of Lemma 7.1 are satisfied and N is a space form $R^m(c)$, then for every point $P \in M$, there exist orthonormal normal vector fields ξ_x defined in a neighborhood U of P such that*

(i) *ξ_x are parallel in the normal bundle on U;*

(ii)
$$A_x = \begin{pmatrix} 0 & 0 & 0 \\ \hline 0 & c_x I_{r_x} & 0 \\ \hline 0 & 0 & 0 \end{pmatrix}$$

where I_{r_x} is the $r_x \times r_x$ identity matrix, the zero matrix in the upper left-hand corner is of degree $r_1 + \cdots + r_{x-1}$ and the A_x's are expressed with respect to their common orthonormal principal directions E_i. Note that $A_x = 0$, if $r_1 + \cdots + r_{x-1} = n$ and we may assume that $A_x = 0$ implies $A_y = 0$ for $y > x$;

(iii) *each c_x is constant on U.*

Proof. Since the normal connection is flat, by Proposition 1.1 there exist orthonormal normal vector fields ξ_x defined in a neighborhood U of P such that ξ_x are parallel in the normal bundle on U. With such a choice of ξ_x, we have

$$\overline{\nabla}_X h^x_{ji} = \nabla_X h^x_{ji}. \tag{7.8}$$

By Lemma 7.1, $|\overline{\nabla}h| = 0$ and thus $\nabla_X A_x = 0$. Hence, the principal curvatures ρ^x_i of A_x are constant. If $\xi'_x = a^y_x \xi_y$, with (a^y_x) an orthogonal matrix with constant entries, then the ξ'_x are also parallel in the normal bundle and $A'_x = a^y_x A_y$, where A'_x is the second fundamental tensor corresponding to ξ'_x. In what follows we will begin with any ξ_x such that $\nabla^\perp \xi_x = 0$ in U and show that there exists an orthogonal matrix (a^y_x) with constant entries such that the second fundamental tensors A'_x with respect to $\xi'_x = a^y_x \xi_y$ have the desired properties (ii). The claim is clearly true if all the A_x's $= 0$ at P. If this is not the case, we distinguish three cases:

(a) all sectional curvatures of $M > 0$ at P,

(b) all sectional curvatures of $M = 0$ at P,

(c) at least one sectional curvature is nonzero at P and at least one sectional curvature is zero at P.

Suppose that ξ_x and U have been chosen such that (i) is satisfied. Thus, each A_x has constant principal curvatures on U.

Case (a). Lemma 7.1 and (7.7) imply that $A_x = c_x I$. We may assume that $c_1 \neq 0$. Let

$$\xi'_1 = \left(\sum_x c_x \xi_x \right) \bigg/ \left(\sum_x c_x^2 \right)^{1/2}$$

and

$$\overline{\xi}_y = (c_1 \xi_y - c_y \xi_1)/(c_1^2 + c_y^2)^{1/2}$$

for $y > 1$. Then $A'_1 = cI$, $c \neq 0$, $\overline{A}_y = 0$ for $y > 1$ and $\overline{\xi}_y \perp \xi'_1$. Use the Gram-Schmidt orthogonalization process on $\overline{\xi}_2, \ldots, \overline{\xi}_{m-n}$ to obtain $\xi'_2, \ldots, \xi'_{m-n}$. Then $A'_y = 0$ for $y > 1$.

Case (b). Let

$$A_x = \begin{pmatrix} \rho^x_1 & & 0 \\ & \ddots & \\ 0 & & \rho^x_n \end{pmatrix}$$

expressed with respect to the common principal directions E_i of the ξ_x's. We may assume that $\rho^1_1 \neq 0$. Let

$$\xi'_1 = \left(\sum_x \rho^x_1 \xi_x \right) \bigg/ \left(\sum_x (\rho^x_1)^2 \right)^{1/2},$$

$$\overline{\xi}_x = (\rho^1_1 \xi_x - \rho^x_1 \xi_1)/(\rho^x_1 \rho^x_1 + \rho^1_1 \rho^1_1)^{1/2}, \quad x > 1.$$

Again, $\bar{\xi}_x \perp \xi_1'$. Use the Gram-Schmidt orthogonalization process on $\bar{\xi}_2, \ldots, \bar{\xi}_{m-n}$ to obtain $\xi_2', \ldots, \xi_{m-n}'$. Then

$$A_x' = \begin{pmatrix} 0 & 0 & \cdots & 0 \\ 0 & * & & \\ \vdots & & \ddots & \\ 0 & & & * \end{pmatrix}, \quad x > 1$$

and $\rho_1'^1 \neq 0$. Thus, we may assume that $\rho_1^x = 0$ for $x > 1$ and $\rho_1^1 \neq 0$. Since $0 = k(\gamma(E_1, E_j)) = \sum_x \rho_1^x \rho_j^x = \rho_1^1 \rho_j^1$ for $j > 1$, we have $\rho_j^1 = 0$ for $j > 1$. If one of the A_x's for $x > 1$, is not zero, we may assume that it is A_2 and apply the above argument to $\xi_2, \ldots, \xi_{m-n}$ and $A_2, \ldots, A_{m-n}$ restricted to the span $\{E_2, \ldots, E_n\}$. We obtain $\rho_j^2 = 0$ for $j > 2$ and $\rho_2^x = 0$ for $x > 2$. It is now clear that an induction argument will work.

Case (c). Order E_i so that $k(\gamma(E_1, E_k)) > 0$ for $2 \leq k \leq r_1$, and $k(\gamma(E_1, E_k)) = 0$, for $k > r_1$. Then Lemma 7.1 and (7.7) imply that $\rho_1^x = \rho_k^x$ for $1 \leq k \leq r_1$. Define ξ_x' as in case (b). We see that we may assume that $\rho_k^x = 0$, for $1 \leq k \leq r_1$; $2 \leq x \leq m - n$. Then

$$k(\gamma(E_1, E_k)) = \rho_1^1 \rho_k^1 = 0, \quad \text{for } k > r_1.$$

Thus, $\rho_k^1 = 0$ for $k > r_1$. If $k(\gamma(E_i, E_j)) \neq 0$ for some $i, j > r_1$, we repeat the above argument applied to $\xi_2, \ldots, \xi_{m-n}$ and $A_2, \ldots, A_{m-n}$ restricted to the span $\{E_{r_1+1}, \ldots, E_n\}$. If $k(\gamma(E_i, E_j)) = 0$ for all $i, j > r_1$, we apply the argument of case (b) to $\xi_2, \ldots, \xi_{m-n}$ and $A_2, \ldots, A_{m-n}$ restricted to the span $\{E_{r_1+1}, \ldots, E_n\}$. In either case we obtain the desired form for A_1 and A_2. It is clear that an induction argument will work. This proves the lemma. $\square$

Now, we consider an example. Let $M^{n_i} = S^{n_i}(s_i)$ be a small (n_i)-sphere of radius s_i immersed in a euclidean $(n_i + 1)$-sphere E^{n_i+1} in a standard way for $i = 1, \ldots, k - 1$. For $n_i = 1$, we assume that the image is a circle; for $n_i \geq 2$, the (isometric) immersion is unique up to an isometry of E^{n_i+1}. Let ξ_i be the inward normal to M^{n_i}. Let $M^{n_k} = E^{n_k}$ and let M^{n_k} be isometrically immersed in $E^{n_k+m-n-k+1}$ such that the image is of the form

$$S^1(r_1) \times \cdots \times S^1(r_t) \times E^{n_k-t},$$

where each $S^1(r_i)$ is a circle of radius r_i in some euclidean plane N_i, $N_i \perp N_j$ for $i \neq j$, and $N_i \perp E^{n_k-t}$. Let ξ_{k+i-1} be the inward normal to $S^1(r_i)$ in N_i and let $\xi_{k+t}, \ldots, \xi_{m-n}$ be normal to E^{n_k-t} and N_i and be constant. Let

$$M = M^{n_1} \times \cdots \times M^{n_k}$$

with the product immersion in E^m. We may consider ξ_x as normal to M in E^m. Let A_x be the corresponding second fundamental tensors. Then the normal connection is flat, the mean curvature vector is parallel in the normal bundle, the length of second fundamental form is constant, all sectional curvatures of $M \geq 0$, all ξ_x are parallel in the normal bundle on M; and the A_x's have the form of (ii) of Lemma 7.2.

Theorem 7.1 (Erbacher, 1972; Yano and Ishihara, 1971). *Let M be an n-dimensional complete submanifold of nonnegative sectional curvature in a euclidean m-space E^m or unit m-sphere $S^m(1)$. Suppose that the mean curvature vector is parallel in the normal bundle and the normal connection is flat. If M is either compact or has constant scalar curvature, then M is the following product submanifold:*

$$M^{n_1} \times \cdots \times M^{n_k},$$

where each M^{n_i} is a small (n_i)-sphere of some radius of E^m or $S^m(1)$; except possibly one of M^{n_i} is a great (n_i)-sphere of E^m or $S^m(1)$. (If this occurs for the latter case, $k = 1$.) The corresponding local result is true with the assumption of constant scalar curvature.

Proof. We may assume the ambient space is euclidean since if the ambient space is $S^m(1)$, we just imbedded $S^m(1)$ as a unit hypersphere of E^{m+1}, then M regarded as a submanifold in E^{m+1} has the properties in the theorem. Let ξ_x be chosen as in Lemma 7.2; we may assume that $c_y \neq 0$ for $1 \leq y \leq k - 1$, and $c_y = 0$ for $y \geq k$ (if all $c_x = 0$, then M is totally geodesic).

Define distributions $T_1, \ldots, T_k$ by

$$T_y(P) = \{X \in T_P(M): A_y(X) = c_y X\}, \quad \text{for } y < k,$$
$$T_k(P) = \{X \in T_P(M): A_x(X) = 0, \quad 1 \leq x \leq m - n\}.$$

Let $n_x = \dim T_x$ (n_k may be zero). Assume that M is connected, simply connected, and complete. Then each T_x is globally defined (for $\xi \in T_P^\perp(M)$, the parallel translate of ξ with respect to the normal connection is independent of path if the normal connection is flat and M is simply connected). Each T_x has constant dimension and is differentiable (the eigenspaces of the A_x have constant dimension, and thus we may find differentiable orthonormal principal directions). The T_x's are orthogonal to each other and

$$T_P(M) = T_1(P) \oplus \cdots \oplus T_k(P) \quad \text{(orthogonal direct sum)}. \tag{7.9}$$

We claim that each T_x is involutive and parallel (that is, for each $Y \in T_x$, $\nabla_X Y \in T_x$, for any vector field X in M).

Since $\nabla A_x = 0$, for any vector fields X, Y in M,

$$0 = (\nabla_X A_x)Y = \nabla_X(A_x(Y)) - A_x(\nabla_X Y).$$

If Y is a principal direction of A_x belonging to the principal curvature c_x, we have

$$c_x \nabla_X Y - A_x(\nabla_X Y) = 0.$$

Thus, $\nabla_X Y$ is in the principal direction of ξ_x with the principal curvature c_x. Thus each T_x is involutive and therefore, from (7.9), each T_x is parallel. This proves the claim.

For each point $P \in M$, let M^{n_x} be the maximal integral submanifold of T_x through P. From the claim, we conclude that M is the Riemannian product $M^{n_1} \times \cdots \times M^{n_k}$ of M^{n_x}. If $n_x = 1$, then $M^{n_x} = \mathcal{R}$ since we have assumed that M is simply connected and complete. If $n_x > 1$, then the curvature tensor of M^{n_x} is the restriction of the curvature tensor of M, since M^{n_x} is totally geodesic (since, for any $X, Y \in T_x$, $\nabla_X Y \in T_x$). Hence, the sectional curvature of M^{n_x} is constant and equal to c_x^2. Also $M^{n_k} = E^{n_k}$. Thus, M is a product of small spheres and possibly a great sphere. Clearly, the corresponding local result is true if we do not assume completeness since we only used completeness to obtain M^{n_x} as the entire sphere.

The second fundamental forms and the normal connection forms of our submanifold with respect to $\xi_1, \ldots, \xi_{m-n}$, chosen as in Lemma 7.2, are the same as those of our example. Thus, by the fundamental theorem of submanifolds (Rigidity Theorem in §2 of Chapter 2), we see that the submanifold M is the product submanifold as in the theorem up to a rigid motion.

If M is complete and connected but not simply connected, let $\overline{M}$ be its simply connected Riemannian covering manifold and let π be the covering map. Then the composition mapping Ψ of $\overline{M}$ in E^m (or E^{m+1}) under π and the immersion of M satisfies the assumptions of the theorem and, by the above, there exists an isometry σ of E^m (or E^{m+1}) such that $\Psi = \sigma \circ \tau$, where τ is the immersion of the example. Thus, M is immersed as the product submanifold of the theorem. This completes the proof of the theorem. □

Remark 7.1. If the codimension is one, then Theorem 7.1 was first obtained by Nomizu and Smyth (1969a).

Remark 7.2. Let M be an n-dimensional submanifold of an m-dimensional space form $R^m(c)$ of curvature c. Let ξ_x be orthonormal normal frames of M. Then we have $[A_1, A_x] = 0$ if ξ_1 is parallel in the normal bundle. By using this fact a method similar to the derivation of formula (7.7), Smyth (1973) proves that

$$\frac{1}{2}\Delta N(A_1) = \sum_{i<j} k(\gamma(E_i, E_j))(\rho_i^1 - \rho_j^1)^2 + |\nabla A_1|^2, \qquad (7.10)$$

where E_i are the principal directions with respect to ξ_1 with the principal curvature ρ_i^1 and $k(\gamma(E_i, E_j))$ denotes the sectional curvature of the plane section $\gamma(E_i, E_j)$. By using this fact, Smyth obtains the following

Theorem 7.2 (Smyth, 1973). *Let M be a closed n-dimensional submanifold of nonnegative sectional curvature with parallel mean curvature vector in a euclidean m-space E^m. Then M is a product submanifold $M_1 \times \cdots \times M_k$, where each M_i is a minimal submanifold in a small hypersphere of a linear subspace of E^m.*

For the proof of this theorem, see Smyth (1973).

Problems

1. Prove that if M is a simply connected n-dimensional submanifold of an m-dimensional space form $R^m(c)$ with flat normal connection and if the normal bundle $T^\perp(M)$ of M in $R^m(c)$ is parallelizable (that is, there exist $m - n$ orthonormal normal sections on M), then there exist $m - n$ orthonormal parallel normal sections on M (defined globally on M).

2. Prove Eqs. (2.5) and (2.6).

3. Prove that if we drop in the axiom of spheres the requirement that M has parallel mean curvature vector, then this weaker axiom implies that M is conformally flat for $m > 3$, and $m > n > 2$ (Schouten, 1954).

4. Prove formula (7.10).

5. Let M be a surface in an m-dimensional space form $R^m(c)$ of curvature c. Prove that if there exists a parallel isoperimetric section on M, then either the normal connection of the normal bundle is flat or M is contained in a hypersphere of $R^m(c)$.

6. Let n be a power of two. Prove that if M is an n-dimensional real projective space imbedded in a $2n$-dimensional space form $R^{2n}(c)$ with constant mean curvature, then M is minimal (Yau, 1974).

7. Let M be an n-dimensional closed submanifold of an m-dimensional space form $R^m(c)$ of curvature c. If the mean curvature vector H is parallel in the normal bundle and the scalar curvature r satisfies the following inequality:

$$r > (n-2)\langle h \rangle^2 + (n-1)(n-2)c,$$

prove that M is contained in an n-sphere of $R^m(c)$ for $n > 2$ (Chen and Okumura, 1973).

Chapter 5

Conformally Flat Submanifolds*

1 Quasiumbilicity

Let M be an n-dimensional manifold immersed in an m-dimensional Riemannian manifold N and ξ a normal section on M. We put $U = \{P \in M: A_\xi$ has at least two distinct eigenvalues at $P\}$. Then U is an open subset of M. Let h_ξ denote the symmetric tensor of type (0,2) associated with A_ξ. If there exist, on U, two functions α, β and a unit 1-form ω such that $h_\xi = \alpha g + \beta \omega \otimes \omega$ on U, then ξ is called a *quasiumbilical section on M* and the submanifold M is said to be *quasiumbilical with respect to ξ*. Since $M - U$ is umbilical in N with respect to ξ, $h_\xi = \alpha g$ for some function α on $M - U$. Thus, if we extend the function β to M with $\beta = 0$ on $M - U$, then we may write h_ξ as follows:

$$h_\xi = \alpha g + \beta \omega \otimes \omega. \tag{1.1}$$

It is clear that α is differentiable on M and β is continuous on M. In particular, if $\alpha = 0$ identically, then ξ is called a *cylindrical section* on M and M is said to be *cylindrical with respect to ξ*; if $\alpha = \beta = 0$ identically, then ξ is called a *geodesic section* on M and M is said to be *geodesic with respect to ξ*; and if $d\alpha$ is nowhere

---

*In this chapter we shall consider only submanifolds of dimension > 2.

zero, then ξ is called a *special quasiumbilical section on M* and *M* is said to be *special quasiumbilical with respect to ξ*.

If there exist locally $m - n$ mutually orthogonal normal vector fields $\xi_1, \ldots, \xi_{m-n}$, such that *M* is quasiumbilical with respect to every normal direction ξ_x; $x = 1, \ldots, m - n$, then the submanifold *M* is said to be *totally quasiumbilical* in *N*.

It is clear that *M* is umbilical with respect to the normal direction ξ if and only if $\beta = 0$ identically, that is, $U = \emptyset$. Moreover, *M* is quasiumbilical with respect to the normal direction ξ if and only if the principal curvatures of ξ are given by $\alpha, \ldots, \alpha, \alpha + \beta$ [α repeats $(n - 1)$-times]. It is easy to verify that umbilicity and quasiumbilicity with respect to one normal direction are conformal invariant properties.

Proposition 1.1 (Chen and Yano, 1972b). *Let M be an n-dimensional totally quasiumbilical submanifold of a conformally flat space N of dimension m. If $n > 3$, then M is conformally flat.*

Proof. Without loss of generality, we may assume that the ambient space *N* is a euclidean *m*-space E^m. Since *M* is totally quasiumbilical, there exist locally $m - n$ mutually orthogonal unit normal vector fields ξ_x such that *M* is quasiumbilical with respect to every normal direction ξ_x. For each fixed x, let

$$h^x = \alpha^x g + \beta^x \omega^x \otimes \omega^x, \tag{1.2}$$

where α^x, β^x are functions in *M* and $\omega^x = \omega_j^x dx^j$ are unit 1-forms in *M*. We put

$$K_{kjix}^h = h_{kx}^h h_{jix} - h_{jx}^h h_{kix}, \tag{1.3}$$

$$K_{jix} = K_{tjix}^t, \quad K_x = g^{ji} K_{jix} \tag{1.4}$$

and

$$L_{jix} = -\frac{K_{jix}}{n-2} + \frac{K_x g_{ji}}{2(n-1)(n-2)}. \tag{1.5}$$

Then, by substituting (1.2) into (1.3), (1.4), and (1.5), we obtain

$$K_{kjix}^h = (\alpha^x)^2 (\delta_k^h g_{ji} - \delta_j^h g_{ki})$$
$$+ \alpha^x \beta^x \{(\delta_k^h \omega_j^x - \delta_j^h \omega_k^x)\omega_i^x - (g_{ji}\omega_k^x - g_{ki}\omega_j^x)\omega^{hx}\}, \tag{1.6}$$

$$K_{jix} = \{(n-1)(\alpha^x)^2 + \alpha^x \beta^x\}g_{ji} + (n-2)\alpha^x \beta^x \omega_j^x \omega_i^x, \tag{1.7}$$

$$K_x = n(n-1)(\alpha^x)^2 + 2(n-1)\alpha^x \beta^x \tag{1.8}$$

and

$$L_{jix} = -\frac{1}{2}(\alpha^x)^2 g_{ji} - \alpha^x \beta^x \omega_j^x \omega_i^x, \tag{1.9}$$

where $\omega^{hx} = \omega_t^x g^{th}$. From (1.6), (1.7), (1.8), and (1.9), we have

$$K_{kjix}^h + \delta_k^h L_{jix} - \delta_j^h L_{kix} + L_{kx}^h g_{ji} - L_{jx}^h g_{ki} = 0, \tag{1.10}$$

where $L_{kx}^h = L_{ktx} g^{th}$. Let C_{kjix}^h denote the left-hand side of Eq. (1.10). Then we have

$$C_{kji}^h = \sum_x C_{kjix}^h.$$

Hence, from (1.10), we see that the conformal curvature tensor C vanishes. From Theorem I.5.1, we know that the submanifold M is conformally flat. This proves the proposition. □

2   Conformally Flat Submanifolds of Codimension 2

Let $\mathfrak{G}_n$ be the set of all real symmetric square matrices of order n and p be a natural number. We put

$$A^{[p]} = \text{trace}(A^p), \quad A \in \mathfrak{G}_n, \tag{2.1}$$

where A^p denotes the product of A with itself p times. For any $A \in \mathfrak{G}_n$, let $\rho_1, \ldots, \rho_n$ denote the eigenvalues of A. We define a real function σ on $\mathfrak{G}_n$ by

$$\sigma(A) = \sum (\rho_k - \rho_j)^2 (\rho_i - \rho_h)^2, \tag{2.2}$$

where the summation is taken over all distinct k, j, i, h. We notice that $\sigma(A) = 0$ if and only if at least $n - 1$ of ρ's are equal.

Lemma 2.1. *For any $A \in \mathfrak{G}_n$ $(n > 3)$, we have*

$$\frac{1}{2}\sigma(A) = -n(n-1)A^{[4]} + 4(n-1)A^{[1]}A^{[3]} + (n^2 - 3n + 3)(A^{[2]})^2$$
$$- 2n(A^{[1]})^2 A^{[2]} + (A^{[1]})^4. \tag{2.3}$$

Proof. By a direct computation, we find that both sides of (2.3) are equal to

$$2(n-2)(n-3)\sum_{k<j} \rho_k^2 \rho_j^2 - 4(n-3)\sum_{i \neq k, j}\sum_{k<j} \rho_k \rho_j \rho_i^2 + 24 \sum_{k<j<i<h} \rho_k \rho_j \rho_i \rho_h.$$

This proves the lemma. □

Lemma 2.2 (Chen and Yano, 1973b). *Let M be a conformally flat space of codimension 2 in a euclidean $(n+2)$-space E^{n+2}. Then we have*

$$\sigma(H) = \sigma(K), \tag{2.4}$$

where $H = A_1$ and $K = A_2$.

Proof. Since the ambient space is euclidean, the Riemann-Christoffel curvature tensor of M has the local expression

$$K^h_{kji} = h^h_{k1}h_{ji1} - h^h_{j1}h_{ki1} + h^h_{k2}h_{ji2} - h^h_{j2}h_{ki2}.$$

Hence, we find

$$K_{ji} = H^{[1]}h_{ji1} - h^1_{jt}h^t_{i1} + K^{[1]}h_{ji2} - h^2_{jt}h^t_{i2}, \tag{2.5}$$

$$r = (H^{[1]})^2 - H^{[2]} + (K^{[1]})^2 - K^{[2]}, \tag{2.6}$$

and

$$L_{ji} = -\frac{H^{[1]}h_{ji1} - h^1_{jt}h^t_{i1} + K^{[1]}h_{ji2} - h^2_{jt}h^t_{i2}}{n-2} + \frac{rg_{ji}}{2(n-1)(n-2)}. \tag{2.7}$$

Substituting (2.7) into (I.5.16) and transvecting the equation obtained by $h^k_{h1}h^{ji}_1$ and $h^k_{h2}h^{ji}_2$, respectively, we obtain

$$\begin{aligned}
h^k_{h1}h^{ji}_1 C^h_{kji} &= \frac{1}{(n-1)(n-2)}[\{-n(n-1)H^{[4]} + 4(n-1)H^{[1]}H^{[3]} \\
&\quad + (n^2 - 3n + 3)(H^{[2]})^2 - 2n(H^{[1]})^2 H^{[2]} + (H^{[1]})^4\} \\
&\quad + \{(n-1)(n-2)((HK)^{[1]})^2 - n(n-1)(HKHK)^{[1]} \\
&\quad - 2(n-1)H^{[1]}K^{[1]}(HK)^{[1]} + 2(n-1)H^{[1]}(HKK)^{[1]} \\
&\quad + 2(n-1)K^{[1]}(HHK)^{[1]} + (H^{[1]})^2(K^{[1]})^2 \\
&\quad - (H^{[1]})^2 K^{[2]} - H^{[2]}(K^{[1]})^2 + H^{[2]}K^{[2]}\}]
\end{aligned}$$

and

$$\begin{aligned}
h^k_{h2}h^{ji}_2 C^h_{kji} &= \frac{1}{(n-1)(n-2)}[\{-n(n-1)K^{[4]} + 4(n-1)K^{[1]}K^{[3]} \\
&\quad + (n^2 - 3n + 3)(K^{[2]})^2 - 2n(K^{[1]})^2 K^{[2]} + (K^{[1]})^4\} \\
&\quad + \{(n-1)(n-2)((KH)^{[1]})^2 - n(n-1)(KHKH)^{[1]} \\
&\quad - 2(n-1)K^{[1]}H^{[1]}(KH)^{[1]} + 2(n-1)K^{[1]}(KHH)^{[1]} \\
&\quad + 2(n-1)H^{[1]}(KKH)^{[1]} + (K^{[1]})^2(H^{[1]})^2 - (K^{[1]})^2 H^{[2]} \\
&\quad - K^{[2]}(H^{[1]})^2 + K^{[2]}H^{[2]}\}].
\end{aligned}$$

Thus, by Lemma 2.1 and the following identities;

$$(HK)^{[1]} = (KH)^{[1]}, \quad (HKHK)^{[1]} = (KHKH)^{[1]},$$

$$(HKK)^{[1]} = (KKH)^{[1]}, \quad (HHK)^{[1]} = (KHH)^{[1]},$$

we obtain

$$0 = (h_{h1}^{k}h_{1}^{ji} - h_{h2}^{k}h_{2}^{ji})C_{kji}^{h}$$
$$= \frac{\sigma(H) - \sigma(K)}{2(n-1)(n-2)}. \tag{2.8}$$

This completes the proof of the lemma. $\square$

Proposition 2.1 (Chen and Yano, 1973b). *Let M be a conformally flat space of codimension 2 in an $(n+2)$-dimensional $(n > 3)$ conformally flat space N. If M is quasiumbilical with respect to one normal direction ξ_1, then it must also be quasiumbilical with respect to the other normal direction ξ_2 orthogonal to ξ_1.*

Proof. Without loss of generality, we may assume that the ambient space N is euclidean. Suppose that M is quasiumbilical with respect to the normal direction ξ_1, then the matrix A_1 has at most two eigenvalues with multiplicity $n - 1$ and 1 or n and 0. Thus, we have $\sigma(A_1) = \sigma(H) = 0$. This implies that the eigenvalues of $K = A_2$ are in the following form

$$x, \dots, x, y \quad (x \text{ repeats } (n-1)\text{-times}),$$

by virtue of Lemma 2.2. This implies that the submanifold M is quasiumbilical with respect to the normal direction ξ_2. This proves the proposition. $\square$

From Proposition 2.1, we have immediately the following

Corollary 2.1 (Cartan, 1917; Schouten, 1921). *Every conformally flat hypersurface of a conformally flat space of dimension > 4 is quasiumbilical.*

Remark 2.1. G.M. Lancaster proved in 1973 that there exist conformally flat hypersurfaces in a euclidean 4-space which are non-quasiumbilical.

3 Special Conformally Flat Spaces

In this section we shall assume that M is an n-dimensional conformally flat hypersurface of an $(n + 1)$-dimensional Riemannian manifold N of constant curvature k and $n > 3$. Hence we have

$$\tilde{K}_{CBA}^{D} = k(\delta_{C}^{D}g_{BA} - \delta_{B}^{D}g_{CA}). \tag{3.1}$$

From Theorem 2.1 we see that the hypersurface M is quasiumbilical in N. Hence on the open subset $U = \{P \in M; h^1 \neq \alpha g \text{ at } p\}$, there exists a unit 1-form $\omega = \omega_i dx^i$ such that

$$h^1 = \alpha g + \beta \omega \otimes \omega, \tag{3.2}$$

where α, β are functions on M with $\beta = 0$ on $M - U$. From (3.1) and (3.2), we find that the curvature tensor K, the Ricci tensor R and the scalar curvature r of the hypersurface M are given, respectively, by

$$K^h_{kji} = (k + \alpha^2)(\delta^h_k g_{ji} - \delta^h_j g_{ki})$$
$$+ \alpha\beta\{(\delta^h_k \omega_j - \delta^h_j \omega_k)\omega_i + (\omega_k g_{ji} - \omega_j g_{ki})\omega^h\}, \tag{3.3}$$
$$K_{ji} = \{(n-1)(k+\alpha^2) + \alpha\beta\}g_{ji} + (n-2)\alpha\beta\omega_j\omega_i, \tag{3.4}$$

and

$$r = n(n-1)(k+\alpha^2) + 2(n-1)\alpha\beta, \tag{3.5}$$

where $\omega^h = \omega_t g^{th}$. Thus we find

$$L = -\frac{1}{2}(k+\alpha^2)g - \alpha\beta\omega \otimes \omega. \tag{3.6}$$

From (3.2) and (3.6) we find

$$L + \alpha h^1 = \frac{1}{2}(\alpha^2 - k)g. \tag{3.7}$$

Since the hypersurface M is conformally flat, $D = 0$, that is,

$$\nabla_k L_{ji} - \nabla_j L_{ki} = 0. \tag{3.8}$$

By the equations (II.2.18) of Codazzi, we have

$$\nabla_k h^1_{ji} - \nabla_j h^1_{ki} = 0. \tag{3.9}$$

Thus, from (3.2), (3.7), (3.8), and (3.9), we find

$$\alpha_k\omega_j - \alpha_j\omega_k = 0 \text{ on } U. \tag{3.10}$$

where $\alpha_k = \partial_k\alpha$, and hence we obtain

$$d\alpha = \gamma\omega \text{ on } U, \tag{3.11}$$

for some function γ on U.

In the following, for a conformally flat space M, if there exist three functions[*] α, β, γ on M such that ω is a unit 1-form on the subset $U = \{P \in M; \beta \neq 0 \text{ at } p\}$ with $d\alpha = \gamma\omega$ on the open subset U and

$$L = -\frac{1}{2}(k+\alpha^2)g - \alpha\beta\omega \otimes \omega \tag{3.12}$$

for some constant k, then M is called a *special conformally flat space* where (3.12) means $L = -\frac{1}{2}(k+\alpha^2)g$ on $M - U$. For a special conformally flat space M,

---

[*]α is differentiable on M and β is continuous on M.

we define a real number $i(M)$ by

$$i(M) = \sup \left\{ k \text{ real}; L = -\frac{1}{2}(k+\alpha^2)g - \alpha\beta\omega \otimes \omega \right.$$

$$\left. \text{for some functions } \alpha \text{ and } \beta \text{ on } M \right\}. \tag{3.13}$$

We call the real number $i(M)$ the *index* of the special conformally flat space M.

From (3.6) and (3.11) we see that if M is a conformally flat hypersurface of an $(n+1)$-dimensional space N of constant curvature k for $n > 3$, then M is a special conformally flat space with index $i(M) \geq k$.

Conversely, if M is a special conformally flat space of dimension n with index i, let k be any constant $\leq i$. Then we have

$$L = -\frac{1}{2}(k+\alpha^2)g - \alpha\beta\omega \otimes \omega \tag{3.14}$$

for some functions α, β, γ on M with $\alpha \geq 0$ where ω is a unit 1-form on the open subset $U = \{P \in M; \beta \neq 0\}$ with $d\alpha = \gamma\omega$ on U. Put

$$h^1 = \alpha g + \beta\omega \otimes \omega. \tag{3.15}$$

Then, from (3.14), we have

$$L = -\alpha h^1 - \frac{1}{2}(k - \alpha^2)g. \tag{3.16}$$

Since M is conformally flat, $C = 0$, or in local components

$$K_{kji}^h = -\delta_k^h L_{ji} + \delta_j^h L_{ki} - L_k^h g_{ji} + L_j^h g_{ki}, \tag{3.17}$$

that is

$$K_{kji}^h = k(\delta_k^h g_{ji} - \delta_j^h g_{ki}) + h_k^{h1} h_{ji}^1 - h_j^{h1} h_{ki}^1 \tag{3.18}$$

On the other hand we have, from (3.16),

$$\nabla_k L_{ji} - \nabla_j L_{ki} = -\alpha_k h_{ji}^1 + \alpha_j h_{ki}^1 - \alpha(\nabla_k h_{ji}^1 - \nabla_j h_{ki}^1) + \alpha\alpha_k g_{ji} - \alpha\alpha_j g_{ki}$$

$$= 0,$$

from which, we find that if $k < i$, then $\alpha^2 > 0$ and we have

$$\nabla_k h_{ji}^1 - \nabla_j h_{ki}^1 = 0, \tag{3.19}$$

by virtue of (3.14) and (3.15). From (3.18), (3.19), and the fundamental theorem of submanifolds, we see that every simply connected special conformally flat space with index i can be isometrically immersed in a space form of curvature $k < i$ as a hypersurface. Consequently, we have the following:

Theorem 3.1 (Chen and Yano, 1972e). *Every conformally flat space M of dimension $n > 3$ of codimension one in a space form is special. Conversely, every simply connected special conformally flat space with index i can be isometrically immersed in any space form of curvature $k < i$ as a hypersurface, and cannot be isometrically immersed in any space form of curvature $k > i$ as a hypersurface.*

For the case $n = 3$, if M is a 3-dimensional quasiumbilical hypersurface in a space form of constant curvature k with the second fundamental tensor given by

$$h^1 = \alpha g + \beta \omega \otimes \omega, \tag{3.20}$$

where α, β are functions on M and ω is a unit 1-form on the open subset $U = \{P \in M; \beta \neq 0 \text{ at } P\}$ with $d\alpha = \gamma \omega$ on U. Then, by a computation similar to the derivation of (3.7), we have

$$L + \alpha h^1 = \frac{1}{2}(\alpha^2 - k)g. \tag{3.21}$$

Hence, by the equation (II.6.11) of Codazzi, we find

$$\nabla_k L_{ji} - \nabla_j L_{ki} = \alpha_j h^1_{ki} - \alpha_k h^1_{ji} + \alpha \alpha_k g_{ji} - \alpha \alpha_j g_{ki}. \tag{3.22}$$

By substituting (3.20) into (3.22), we find

$$\nabla_k L_{ji} - \nabla_j L_{ki} = 0. \tag{3.23}$$

Therefore, by (I.5.21) and (3.23), we see that the Cotton tensor D vanishes identically. Hence, by Theorem I.5.1, we know that M is conformally flat. Consequently, we have

Theorem 3.2 (Chen and Yano, 1972e). *Let M be a quasiumbilical hypersurface of a 4-dimensional space form with the second fundamental form having the form (3.20), then M is conformally flat.*

Remark 3.1. It came to my notice when I had already finished this book that Kulkarni obtained very recently some similar results on conformally flat hypersurfaces which are obtained by Yano and the author, see Kulkarni (1972).

4 Locus of Spheres

In the following, an n-dimensional submanifold M in an m-dimensional space form $R^m(k)$ is called a locus of r-spheres if M is obtained by the smooth glueing of some n-dimensional submanifolds of M (possibly with boundary) such that each of the submanifolds is foliated by r-spheres $(0 < r < n)$ of $R^m(k)$.

Theorem 4.1 (Chen and Yano, 1972c). *A special quasiumbilical hypersurface M of an $(n+1)$-dimensional space form $R^{n+1}(k)$ is conformally flat for $n > 3$ and it is foliated by $(n-1)$-spheres of $R^{n+1}(k)$.*

Proof. Let M be a special quasiumbilical hypersurface of an $(n+1)$-dimensional space form $R^{n+1}(k)$. Then we have

$$h^1 = \alpha g + \beta \omega \otimes \omega \tag{4.1}$$

with

$$d\alpha \neq 0 \text{ everywhere} \tag{4.2}$$

for some functions α, β and a unit 1-form $\omega = \omega_i dx^i$ on the subset $\{P \in M;\ \beta \neq 0$ at $P\}$. By Proposition 1.1, we see that M is conformally flat for $n > 3$. Hence, the Cotton tensor D vanishes identically, or in local components

$$\nabla_k L_{ji} - \nabla_j L_{ki} = 0. \tag{4.3}$$

On the other hand, we have, from (3.6) and (4.1), that

$$L + \alpha h^1 = \frac{1}{2}(\alpha^2 - k)g. \tag{4.4}$$

Hence, by (4.3), (4.4), and the equation (II.6.11) of Codazzi, we find

$$\beta\{\alpha_k \omega_j - \alpha_j \omega_k\} = 0, \tag{4.5}$$

where the α_k are given by $d\alpha = \alpha_k dx^k$. Combining (4.2) and (4.5), we have

$$h^1 = \alpha g + \gamma\, d\alpha \otimes d\alpha, \quad d\alpha \neq 0 \text{ everywhere,} \tag{4.6}$$

for some function γ. Substituting (4.6) into the equation (II.6.11) of Codazzi, we obtain

$$\alpha_k g_{ji} - \alpha_j g_{ki} + \gamma_k \alpha_j \alpha_i - \gamma_j \alpha_k \alpha_i + \gamma \alpha_j (\nabla_k \alpha_i) - \gamma \alpha_k (\nabla_j \alpha_i) = 0, \tag{4.7}$$

where the γ_k are given by $d\gamma = \gamma_k dx^k$. Transvecting (4.7) by α^k, we find

$$(\alpha_t \alpha^t)g_{ji} + (\gamma_t \alpha^t - 1)\alpha_j \alpha_i - (\alpha_t \alpha^t)\gamma_i \alpha_i + \gamma \alpha_j (\alpha^t \nabla_t \alpha_i) - \gamma(\alpha_t \alpha^t)\nabla_j \alpha_i = 0. \tag{4.8}$$

Equation (4.8) shows that $\gamma \nabla_j \alpha_i$ is of the form;

$$\gamma \nabla_j \alpha_i = g_{ji} + q_j \alpha_i + q_i \alpha_j + \frac{\gamma_t \alpha^t - 1}{\alpha_t \alpha^t} \alpha_j \alpha_i, \tag{4.9}$$

where q_j defines a 1-form $q_j dx^j$ on M.

Since $\alpha_i = \partial_i \alpha$ and $d\alpha \neq 0$ everywhere, the differential system

$$\Omega = \{f\,d\alpha;\ f \text{ function on } M\}$$

defines a foliation on M and $\alpha = \text{constant}$ defines a family of maximal integrable hypersurfaces in M. We represent one of them, V, locally by

$$x^h = x^h(\zeta^a),$$

where $\{\zeta^a; a = 1, \ldots, n-1\}$ denotes a local coordinate system in V.
We put

$$\partial_b = \partial/\partial\zeta^b, \quad N^h = \alpha^h/(\alpha_t\alpha^t)^{1/2}, \quad \alpha^h = \alpha_t g^{th} \tag{4.10}$$

and let h' denote the second fundamental form of V in M. Then ∂_b are the basis vector fields in V and can be regarded as vector fields in M and $R^{n+1}(k)$. Since $N = N^h \partial_h$, $\partial_h = \partial/\partial x^h$, is a unit normal vector field of V in M, we have

$$g(N, \partial_b) = 0, \tag{4.11}$$

from which, we find

$$\gamma(\alpha_t\alpha^t)^{1/2} h'(X', Y') = g'(X', Y')N \tag{4.12}$$

by virtue of (4.9), for any vector fields X', Y' in V, where g' is the induced metric tensor on V. Equation (4.12) implies that γ never vanishes and

$$h'(X', Y') = \frac{g'(X', Y')}{\gamma(\alpha_t\alpha^t)^{1/2}} N. \tag{4.13}$$

On the other hand, from (4.1) and (4.10), we find

$$h(X', Y') = \alpha g(X', Y')\xi \tag{4.14}$$

for any vector fields X', Y' in V, where ξ is a unit normal vector field of M in $R^{n+1}(k)$. Thus, by combining (III.3.4), (4.13) and (4.14), we see that the second fundamental form $\bar{h}$ of V in $R^{n+1}(k)$ satisfies

$$\bar{h}(X', Y') = \alpha g'(X', Y')\xi + \frac{g'(X', Y')}{\gamma(\alpha_t\alpha^t)^{1/2}} N \tag{4.15}$$

for any vector fields X', Y' in V. Hence, V is a totally umbilical submanifold of $R^{n+1}(k)$. Consequently, by Proposition II.3.2, we see that V is contained in an $(n-1)$-sphere of $R^{n+1}(k)$. This shows that the hypersurface M is foliated by $(n-1)$-spheres of $R^{n+1}(k)$. This proves the theorem. $\qquad\square$

Remark 4.1. If $n = 3$ and M is a conformally flat hypersurface in a 4-dimensional space form $R^4(k)$ of curvature k, then, by a proof similar to that of Theorem 4.1, we may show that M is a locus of 2-spheres whenever M is a special quasiumbilical hypersurface in $R^4(k)$.

Theorem 4.2 (Chen and Yano, 1972c). *A locus of $(n-1)$-spheres $(n > 3)$ in an $(n+1)$-dimensional space form $R^{n+1}(k)$ of curvature k is a conformally flat space if and only if the unit normal vector field of the hypersurface M in $R^{n+1}(k)$, restricted to an $(n-1)$-sphere V, is parallel with respect to the normal bundle of the $(n-1)$-sphere V in $R^{n+1}(k)$.*

Proof. Suppose that M is a locus of $(n-1)$-spheres in an $(n+1)$-dimensional space form $R^{n+1}(k)$ of curvature k with ξ as a unit normal vector field on M. Let V be one of the $(n-1)$-spheres and η the unit normal vector field of V in M, and we also denote by ξ the restriction of ξ on V. Then, by Eq. (III.3.4), we see that the second fundamental forms $h', \bar{h}$ of V in M and $R^{n+1}(k)$, respectively, satisfy

$$\bar{h}(X', Y') = h'(X', Y') + h(X', Y') \tag{4.16}$$

for any vector fields X', Y' in V, where h is the second fundamental form of M in $R^{n+1}(k)$.

Let $\tilde{\nabla}$ denote the connection on $R^{n+1}(k)$. Then for any vector field X' in V, we have, from Weingarten's formula,

$$\begin{aligned}
\tilde{\nabla}_{X'}\xi &= -A'_\xi(X') + \nabla'^\perp_{X'}\xi \\
&= -A'_\xi(X') + \theta(X')\eta,
\end{aligned} \tag{4.17}$$

where $\nabla'^\perp$ is the normal connection of V in $R^{n+1}(k)$, $\theta(X')\eta = \nabla'^\perp_{X'}\xi$, and A'_ξ is the second fundamental tensor of V in $R^{n+1}(k)$ with respect to ξ.

Since, we have, from Gauss' formula,

$$\tilde{\nabla}_{X'}\eta = \nabla_{X'}\eta + h(X', \eta) \tag{4.18}$$

and, from Weingarten's formula,

$$\tilde{\nabla}_{X'}\eta = -A'_\eta X' - \theta(X')\xi, \tag{4.19}$$

for any vector field X' in V, we find

$$h(X', \eta) = -\theta(X')\xi. \tag{4.20}$$

Let X, Y be any two vector fields in M defined along V, we decompose X, Y into

$$X = X' + \tilde{g}(X, \eta)\eta, \quad Y = Y' + \tilde{g}(Y, \eta)\eta, \tag{4.21}$$

where X' and Y' are the tangential components of X and Y on V. By using (4.20), we find

$$h(X, Y) = h(X', Y') - \{\theta(X')\tilde{g}(Y, \eta) + \theta(Y')\tilde{g}(X, \eta)\}\xi + \tilde{g}(X, \eta)\tilde{g}(Y, \eta)\rho\xi, \tag{4.22}$$

where $\rho\xi = h(\eta, \eta)$.

Since, V is an $(n-1)$-sphere in $R^{n+1}(k)$, V is totally umbilical in $R^{n+1}(k)$. Hence, from (4.16), we find

$$h(X', Y') = \sigma g(X', Y')\xi, \tag{4.23}$$

for some function σ on V. Combining (4.22) and (4.23), we find

$$h(X, Y) = \{\sigma g(X', Y') - [\theta(X')\tilde{g}(Y, \eta) + \theta(Y')\tilde{g}(X, \eta)] + \tilde{g}(X, \eta)\tilde{g}(Y, \eta)\rho\}\xi. \tag{4.24}$$

Hence, from Proposition 1.1 and Theorem 2.1, we see that M is conformally flat if and only if there exists a function f such that

$$\theta(X') = f\tilde{g}(X, \eta) \tag{4.25}$$

for any vector field X in M defined along V. It is clear that (4.25) holds when and only when $\theta = 0$ and $\theta = 0$ when and only when ξ is parallel in the normal bundle of V in $R^{n+1}(k)$. Thus, M is conformally flat if and only if ξ is parallel in the normal bundle of V in $R^{n+1}(k)$. This proves the theorem. $\qquad\square$

Theorem 4.3. *Let M be a quasiumbilical hypersurface of an $(n+1)$-dimensional space form $R^{n+1}(k)$ with the second fundamental form h^1 given by $\alpha g + \beta \omega \otimes \omega$ such that α is constant and β is nowhere zero. Then M is foliated by $(n-1)$-spheres such that each of the $(n-1)$-spheres is of the same constant curvature $k + \alpha^2$.*

Under the hypothesis of this theorem, the distributions $\mathcal{D} = \{X_P \in T_P(M);$ $\omega(X_P) = 0, P \in M\}$ and $\mathcal{D}^1 = \{Y_P \in T_P(M); Y_P \perp \mathcal{D}_P, P \in M\}$ are both involutive and, by the Codazzi equation, we have

$$\nabla_j \omega_i = (\omega^k \beta_k)\omega_j \omega_i - \beta_j \omega_i + \beta(\nabla_k \omega_j)\omega^k \omega_i - \beta(\nabla_k \omega_i)\omega^k \omega_j, \tag{4.26}$$

where $\beta_k = \nabla_k \beta$. By using this fact we may prove Theorem 4.3 in exactly the same way as in the proof of Theorem 4.1.

Combining Theorem 4.1, Theorem 4.3, and Proposition II.3.2, we have the following

Theorem 4.4. *Let M be a conformally flat hypersurface of an $(n+1)$-dimensional space form $R^{n+1}(k)$ with $n > 3$. Then M is locus of $(n-1)$-spheres.*

5 Canal Hypersurfaces

In this section, we shall give an example of a locus of $(n-1)$-spheres which is conformally flat and special quasiumbilical.

In the following, by a *canal hypersurface* we mean the envelope of a one-parameter family of small hyperspheres of a euclidean $(n+1)$-space E^{n+1}.

Let M be a canal hypersurface in E^{n+1} given by the envelope of the following one-parameter family of hyperspheres:

$$(\tilde{X} - x(s)) \cdot (\tilde{X} - x(s)) = r(s)^2, \quad r(s) > 0, \tag{5.1}$$

where $\tilde{X}$ denotes the position vector field of E^{n+1} with respect to the origin of E^{n+1}, $x(s)$ and $r(s)$ are, respectively, the centers and the radii of the hyperspheres, and the dot "$\cdot$" denotes the scalar product in E^{n+1}. Then the canal hypersurface M is given by (5.1) and

$$(\tilde{X} - x(s)) \cdot x'(s) = -r(s)r'(s), \tag{5.2}$$

where $x' = dx/ds$ and $r' = dr/ds$. Without loss of generality, we may assume that the canal hypersurface M is also given by a vector function in E^{n+1}:

$$\tilde{X} = \tilde{X}(u^1, \ldots, u^{n-1}, s) \tag{5.3}$$

satisfying (5.1) and (5.2), where $\{u^1, \ldots, u^{n-1}, u^n = s\}$ is a local coordinate system in M. By taking partial derivatives of (5.1), we obtain

$$X_b \cdot (\tilde{X} - x(s)) = 0, \quad b = 1, \ldots, n-1, \tag{5.4}$$

and

$$X_n \cdot (\tilde{X} - x(s)) = 0, \tag{5.5}$$

by virtue of (5.2), where $X_b = \partial\tilde{X}/\partial u^b$ and $X_n = \partial\tilde{X}/\partial s$. From (5.4) and (5.5), we see that the unit normal vector field ξ of M in E^{n+1} is parallel to the vector field $\tilde{X} - x(s)$. Thus, we may write

$$\tilde{X} = x(s) - r(s)\xi(u^1, \ldots, u^{n-1}, s). \tag{5.6}$$

By taking the partial derivative of (5.2) with respect to u^b, we have

$$X_b \cdot x'(s) = 0. \tag{5.7}$$

Since ξ is a unit normal vector field, we have, from Weingarten's formula,

$$\tilde{\nabla}_{X_b}\xi = -A_\xi(X_b) = -h_b^a X_a - h_b^n X_n \tag{5.8}$$

and

$$\tilde{\nabla}_{X_n}\xi = -h_n^a X_a - h_n^n X_n. \tag{5.9}$$

From (5.6) and (5.8), we find

$$X_b = rh_b^a X_a + rh_b^n X_n, \tag{5.10}$$

from which

$$h_b^a = \frac{1}{r}\delta_b^a, \quad h_b^n = 0. \tag{5.11}$$

Also, from (5.6) and (5.9), we have

$$X_n = x' - r'\xi + rh_n^a X_a + rh_n^n X_n. \tag{5.12}$$

Equations (5.7) and (5.12) imply

$$g_{bn} = rh_n^a g_{ba} + rh_n^n g_{bn} = rh_n^i g_{ib}$$
$$= rh_{bn}. \tag{5.13}$$

Thus, from (5.11) and (5.13), we obtain

$$h_{ci} = \frac{1}{r} g_{ci}. \tag{5.14}$$

Hence, if we put

$$\alpha' = \frac{1}{r}, \quad \beta' = h_{nn} - \frac{1}{r} g_{nn}, \tag{5.15}$$

then we obtain

$$h^1 = \alpha' g + \beta'(ds) \otimes (ds), \tag{5.16}$$

where α' is a function of s. Therefore, by Proposition 1.1, M is a conformally flat space. In particular, if we choose the function $r = r(s)$ such that $dr \neq 0$ everywhere, then $d\alpha'$ is nowhere zero. In this case, the canal hypersurface M is special quasiumbilical.

Consequently, we have the following:

Theorem 5.1 (Cartan, 1917: Chen and Yano, 1973a). *Every canal hypersurface M in a euclidean $(n+1)$-space E^{n+1} is conformally flat for $n > 3$ and it is a special quasiumbilical hypersurface if dr is nowhere zero for $n > 2$, where r is defined by (5.1).*

Problems

1. Prove that quasiumbilicity and umbilicity with respect to one normal direction are conformal invariant properties.

2. For a conformally flat space M, let L denote the tensor field given by (I.5.9). If there exist two functions α and β such that $\beta = \beta(\alpha)$ and

$$L = \alpha g + \beta(d\alpha) \otimes (d\alpha),$$

then the conformally flat space M is called a *subprojective space*. Prove that if a canal hypersurface in a euclidean $(n+1)$-space E^{n+1} $(n > 3)$ is subprojective with respect to the induced conformal structure, then the canal hypersurface M is a surface of revolution, that is, the locus of centers lies on a straight line of E^{n+1} (Chen and Yano, 1973a).

3. Let M be a quasiumbilical hypersurface in an $(n + 1)$-dimensional conformally flat space N $(n > 3)$ with the second fundamental tensor h^1 given by

$$h^1 = \alpha g + \beta \omega \otimes \omega$$

where α, β are two functions on M and ω is a unit 1-form on the open subset $U = \{P \in M; \beta \neq 0 \text{ at } P\}$. Let $\tilde{L}$ denote the tensor (I.5.9) for N. If $\tilde{L}$ satisfies

$$\tilde{L}(X, Y) = -\frac{1}{2}(k + \alpha^2)g(X, Y) + \alpha h^1(X, Y)$$

for any vector fields X, Y on M, then M is called a *quasiumbilical hypersurface of type k*. Prove that a hypersurface of an $(n + 1)$-dimensional conformally flat space is of constant curvature k if and only if it is a quasiumbilical hypersurface of type k (Chen and Yano, 1972d).

4. Let M be an n-dimensional Riemannian manifold $(n > 3)$, α and β two functions on M, and $\omega = \omega_i dx^i$ a unit 1-form on M. Prove that if the curvature tensor K of M has the following form:

$$K^h_{kji} = \alpha(\delta^h_k g_{ji} - \delta^h_j g_{ki})$$
$$+ \beta\{(\delta^h_k \omega_j - \delta^h_j \omega_k)\omega_i + (\omega_k g_{ji} - \omega_j g_{ki})\omega^h\}, \quad \omega^h = g^{th}\omega_t,$$

then M is a conformally flat space and it has the following three properties:

(I) The curvature operator $K^h_{kji} X^k Y^j$ associated with two vector fields X and Y orthogonal to the vector field $Z = \omega^i \partial_i$ associated with the unit 1-form ω annihilates Z, that is,

$$K(X, Y; Z, \cdot) = 0.$$

(II) Sectional curvature with respect to a section containing the vector field Z is a constant.

(III) Sectional curvature with respect to a section orthogonal to Z is a constant.

5. Prove that if an n-dimensional Riemannian manifold M $(n > 3)$ satisfies properties (I), (II), and (III), then the curvature tensor K of M has the form given in Problem 4 (Yano, Houh, and Chen, 1973).

6. Prove Theorem 4.3.

7. Prove that a locus of $(n - 1)$-spheres $(n > 3)$ M in an m-dimensional space form $R^m(k)$ curvature of k is a conformally flat space if the unit normal vector field of every leaf in M is parallel in the normal bundle of the leaf in $R^m(k)$, where a leaf in M means a maximal integral submanifold of the foliation on M.

Chapter 6

Umbilical Submanifolds

1 Ricci Curvature and Scalar Curvature for Pseudoumbilical Submanifolds

Let M be an n-dimensional submanifold of an m-dimensional Riemannian manifold N and let Z be a unit vector of M at a point $P \in M$. If, for any vector X in M and any normal vector ξ orthogonal to the mean curvature H of M at P, we have

$$\tilde{g}(Z, A_\xi(X)) = 0, \tag{1.1}$$

then the unit vector Z is called a *Ricci minimal direction* of M in N.

The main purpose of this section is to study the Ricci curvature and the scalar curvature for a pseudoumbilical submanifold in a space of constant curvature.

Proposition 1.1 (Chen and Yano, 1972a). *Let M be an n-dimensional pseudoumbilical submanifold of an m-dimensional Riemannian manifold N of constant curvature k. Then for any unit vector field Z of M, the Ricci curvature $R(Z, Z)$ with respect to Z satisfies*

$$R(Z, Z) \leqq (n-1)(k + |H|^2). \tag{1.2}$$

The equality sign of (1.2) holds when and only when Z is a Ricci minimal direction.

Proof. Suppose that M is an n-dimensional pseudoumbilical submanifold of an m-dimensional Riemannian manifold N of constant curvature k. Then the equation of Gauss is given by

$$K(X, Y; U, V) = k\{g(X, V)g(Y, U) - g(X, U)g(Y, V)\}$$
$$+ \tilde{g}(h(X, V), h(Y, U)) - \tilde{g}(h(X, U), h(Y, V)), \qquad (1.3)$$

or in local components

$$K_{kjih} = k(g_{kh}g_{ji} - g_{jh}g_{ki}) + h^x_{kh}h_{jix} - h^x_{jh}h_{kix}. \qquad (1.4)$$

For simplicity, we may choose $m - n$ mutually orthogonal unit normal vector fields ξ_x such that ξ_1 is in the direction of the mean curvature vector, that is,

$$H = |H|\xi_1. \qquad (1.5)$$

From (II.1.18), we find

$$h^{(1)}h^1_{ji} = (h^{(1)})^2 g_{ji}, \qquad (1.6)$$

where $h^{(1)} = (1/n)h^t_{t1}$.

Transvecting g^{kh} to (1.4) and applying (1.6), we find

$$K_{ji} = (n - 1)(k + |H|^2)g_{ji} - \sum_{y=2}^{m-n} h^y_{ti}h^t_{jy}. \qquad (1.7)$$

Hence, if $Z = Z^i \partial_i$ is any unit vector field in M, then, from (1.7), we see that the Ricci curvature $R(Z, Z)$ with respect to Z is given by

$$R(Z, Z) = K_{ji}Z^j Z^i = (n - 1)(k + |H|^2) - \sum_{y=2}^{m-n} (h^y_{ti}z^i)(h^t_{jy}Z^j). \qquad (1.8)$$

From this we obtain (1.2).

Now, suppose that the equality sign of (1.2) holds, then, by (1.8), we find

$$Z^t h^y_{jt} = 0, \quad j = 1, \ldots, n; \, y = 2, \ldots, m - n. \qquad (1.9)$$

From this we see that, for any vector field X in M and any normal vector field ξ orthogonal to the mean curvature vector H, we have $g(Z, A_\xi(X)) = 0$. Hence, Z is a Ricci minimal direction of M in N. Conversely, if Z is a Ricci minimal direction of M in N, then, by (1.1), we find (1.9). Hence, from (1.8), we see that the equality sign of (1.2) holds. This proves the proposition.   □

Proposition 1.2 (Chen and Yano, 1972a). *Let M be an n-dimensional pseudoumbilical submanifold of an m-dimensional space form $R^m(k)$ of curvature k. Then the scalar curvature r satisfies*

$$r \leqq n(n - 1)(k + |H|^2). \qquad (1.10)$$

The equality sign of (1.10) holds when and only when M is contained in an n-sphere of $R^m(k)$.

Proof. Transvecting (1.7) by g^{ji}, we find

$$r = n(n-1)(k + |H|^2) - \sum_{y=2}^{m-n} h_{ts}^y h_y^{st}. \tag{1.11}$$

This implies (1.10).

If the equality sign of (1.10) holds, then, by (1.11), we find

$$h_{ji}^y = 0, \quad y = 2, \ldots, m-n; i, j = 1, \ldots, n. \tag{1.12}$$

Hence, the open subset $U = \{P \in M; H \neq 0 \text{ at } P\}$ is totally umbilical in $R^m(k)$. Combining this and Proposition II.3.2, we see that each component of U is contained in a small n-sphere of $R^m(k)$. In particular, we see that U is also a closed subset of M. If U is nonempty, then we have $U = M$. Thus, M is contained in a small n-sphere of $R^m(k)$. If U is empty, then M is a minimal submanifold of $R^m(k)$. In this case we have

$$r = n(n-1)k. \tag{1.13}$$

On the other hand, by transvecting g^{kh} to (1.4) and applying the minimality of M in $R^m(k)$, we find

$$K_{ji} = (n-1)kg_{ji} - h_{ti}^x h_{jx}^t. \tag{1.14}$$

By transvecting g^{ji} to (1.14) and applying (1.13), we find

$$h_{ts}^x h_x^{ts} = 0.$$

From this we see that M is totally geodesic in $R^m(k)$. Thus, M is contained in a great n-sphere of $R^m(k)$. The converse of this is trivial. This proves the proposition. □

2 Quasiparallel Normal Directions

Let M be an n-dimensional submanifold of an m-dimensional Riemannian manifold N. For a unit normal vector field ξ of M in N, ξ is said to be *nonparallel* if $\nabla^\perp \xi$ is nowhere zero, where $\nabla^\perp$ is the induced normal connection on the normal bundle $T^\perp(M)$. If η is another unit normal vector field of M in N such that $\nabla^\perp \xi$ has no component in every normal direction orthogonal to η, then the normal direction ξ is said to be *quasiparallel* with respect to the normal direction η. It is clear that if the codimension is two, then every normal direction is quasiparallel with respect to the other normal direction orthogonal to it.

Proposition 2.1 (Chen and Yano, 1973f). *Let M be an n-dimensional submanifold of an m-dimensional Riemannian manifold N of constant curvature k. If M is* *umbilical with respect to a nonparallel normal direction ξ and ξ is quasiparallel with respect to a normal direction η orthogonal to ξ, then the submanifold M is quasiumbilical with respect to the normal direction η.*

Proof. Since ξ and η are two mutually orthogonal unit normal vector fields of M in N, we may choose locally $m - n$ mutually orthogonal unit normal vector fields ξ_x such that $\xi_1 = \xi$ and $\xi_{m-n} = \eta$.

Since M is umbilical with respect to the normal direction ξ and ξ is nonparallel and quasiparallel with respect to η, we have

$$h^1(X, Y) = \alpha g(X, Y), \quad \nabla^\perp \xi = \theta \eta, \tag{2.1}$$

where θ is a nowhere zero 1-form on M and α is a function on M. Hence, from the equation (II.2.18) of Codazzi, we find

$$(X\alpha)g(Y, Z) - (Y\alpha)g(X, Z) - h^{m-n}(Y, Z)\theta(X) + h^{m-n}(X, Z)\theta(Y) = 0, \tag{2.2}$$

for any vector fields X, Y, Z in M, or in local components

$$\alpha_k g_{ji} - \alpha_j g_{ki} - \theta_k k_{ji} + \theta_j k_{ki} = 0, \tag{2.3}$$

where $\alpha_k = \partial_k \alpha$, $\theta_k = \theta(X_k)$, and $k_{ji} = h_{ji}^{m-n}$.

Transvecting $\theta^k = g^{kt}\theta_t$ to (2.3), we get

$$(\alpha_k + k_{kt}\theta^t)\theta_j = (\alpha_j + k_{jt}\theta^t)\theta_k,$$

from which, by transvecting θ^k, we find

$$(\alpha_t\theta^t + k(\theta, \theta))\theta_j = (\alpha_j + k_{jt}\theta^t)|\theta|^2,$$

where $k(\theta, \theta) = k_{ts}\theta^t\theta^s$, $|\theta|^2 = \theta_t\theta^t$.

Consequently, we find

$$\alpha_j + k_{jt}\theta^t = \frac{\alpha_t\theta^t + k(\theta, \theta)}{|\theta|^2}\theta_j. \tag{2.4}$$

Transvecting g^{ki} to (2.3), we find

$$\alpha_j + k_{jt}\theta^t = -(n - 2)\alpha_j + k_t^t\theta_j, \tag{2.5}$$

from which, by transvecting θ^j, we obtain

$$(n - 1)\alpha_t\theta^t + k(\theta, \theta) = k_t^t|\theta|^2. \tag{2.6}$$

From (2.4) and (2.5), we find

$$\frac{\alpha_t\theta^t + k(\theta, \theta) - k_t^t|\theta|^2}{|\theta|^2}\theta_j = -(n - 2)\alpha_j,$$

which, by using (2.6), can be written as

$$\alpha_j = \frac{\alpha_t \theta^t}{|\theta|^2} \theta_j. \tag{2.7}$$

From (2.5) and (2.7), we find

$$\begin{aligned}
k_{jt}\theta^t &= -(n-1)\alpha_j + k_t^t \theta_j \\
&= \frac{-(n-1)\alpha_t \theta^t + k_t^t |\theta|^2}{|\theta|^2} \theta_j,
\end{aligned}$$

from which, by using (2.6), we get

$$k_{jt}\theta^t = \frac{k(\theta, \theta)}{|\theta|^2} \theta_j. \tag{2.8}$$

Transvecting θ^k to (2.3), we find

$$\alpha_t \theta^t g_{ji} - \alpha_j \theta_i - |\theta|^2 k_{ji} + \theta_j k_{it}\theta^t = 0. \tag{2.9}$$

Substituting (2.7) and (2.8) into (2.9), we find

$$k_{ji} = \lambda g_{ji} + \mu \theta_j \theta_i \tag{2.10}$$

where

$$\lambda = \frac{\alpha_t \theta^t}{|\theta|^2}, \quad \mu = \frac{k_t^t - n\lambda}{|\theta|^2}, \tag{2.11}$$

that is,

$$h^{m-n} = \lambda g + \mu \theta \otimes \theta. \tag{2.12}$$

This proves the proposition. □

Proposition 2.2 (Chen and Yano, 1973f). *Let M be an n-dimensional submanifold of an m-dimensional Riemannian manifold of constant curvature k. If M is umbilical with respect to a nonparallel normal direction ξ and ξ is quasiparallel with respect to a normal direction η orthogonal to ξ, then M is cylindrical with respect to the normal direction η if and only if trace $A_\xi = $ constant.*

Proof. Under the hypothesis of the proposition, if we have

$$\text{trace } A_\xi = \text{constant}, \tag{2.13}$$

then, by (2.11), we have $\lambda = 0$. Hence, by (2.12), we see that the submanifold M is cylindrical with respect to the normal direction η.

Conversely, if M is cylindrical with respect to the normal direction η, then, there exist on M, a function β and a 1-form ω such that

$$h^{m-n} = \beta\omega \otimes \omega. \qquad (2.14)$$

Comparing (2.12) and (2.14), we find

$$\gamma\theta = \omega \qquad (2.15)$$

for some function γ on M. This implies that

$$h^{m-n} = f\theta \otimes \theta \qquad (2.16)$$

for some function f on M. Substituting this into (2.2), we find

$$(X\alpha)g(Y, Z) - (Y\alpha)g(X, Z) = 0. \qquad (2.17)$$

Since this is true for any vector field X, Y, Z in M, we obtain $X\alpha = 0$. This implies that α is a constant. Hence, we see that trace $A_\xi = n\alpha$ is a constant. This completes the proof of the proposition. $\square$

From Proposition 2.2, we have

Proposition 2.3 (Chen and Yano, 1972a: Chen, 1971e). *Let M be a pseudo-umbilical submanifold of an m-dimensional Riemannian manifold N of constant curvature k and let ξ be the unit normal direction in the direction of mean curvature vector H. If ξ is nonparallel and quasiparallel with respect to a normal direction η orthogonal to ξ, then M has constant mean curvature if and only if the submanifold M is geodesic with respect to the normal direction η.*

Proof. Since η is perpendicular to the mean curvature vector H, we have trace $A_\eta = 0$. Hence, the proposition follows immediately from Proposition 2.2. $\square$

If the codimension is two, then, from Proposition 2.3, we have immediately the following

Proposition 2.4 (Chen, 1971e; Chen and Yano, 1972a). *Let M be a pseudo-umbilical submanifold of codimension 2 in an m-dimensional space form $R^m(k)$ of curvature k. Then the submanifold M has constant mean curvature if and only if the mean curvature vector is parallel.*

Combining Problem II.7 and Proposition 2.4, we obtain immediately the following:

Proposition 2.5 (Chen, 1971e; Chen and Yano, 1972a). *Let M be a pseudoumbilical submanifold of codimension 2 in an $(n + 2)$-dimensional space form $R^{n+2}(c)$ of curvature c. If M has constant mean curvature, then M is either a minimal submanifold of $R^{n+2}(c)$ or a minimal hypersurface of a small hypersphere of $R^{n+2}(c)$.*

3 Nonparallel Normal Subbundles

Let M be an n-dimensional submanifold of an m-dimensional Riemannian manifold N of constant curvature c. By a *normal q-subbundle* of M in N, we mean a q-plane subbundle of the normal bundle $T^\perp(M)$ of M in N. If M is umbilical with respect to every normal direction in a q-subbundle V, then the submanifold M is said to be *umbilical with respect to the normal subbundle V*. If the covariant derivative of every unit normal direction in V has no component in the complementary normal subbundle $V^\perp$ orthogonal to V, then the subbundle V is said to be *parallel in the normal bundle*, or simply, *parallel*. If, there exists, in V, a normal vector field ξ such that the covariant derivative of ξ has nonzero component everywhere in the complementary normal subbundle $V^\perp$ orthogonal to V, then the normal subbundle V is said to be *nonparallel in the normal bundle*, or just *nonparallel*. For a unit normal vector field ξ of M in N, it is clear that the normal line subbundle generated by ξ is parallel (respectively, nonparallel) if and only if the unit normal vector field ξ is parallel (respectively, nonparallel).

It is also clear that a normal subbundle V is parallel (respectively, nonparallel) if and only if the complementary normal subbundle $\nabla^\perp$ orthogonal to V is parallel (respectively, nonparallel).

In this section, we shall assume that M is umbilical with respect to a nonparallel normal $(m - n - 1)$-subbundle V. In the following, we choose $m - n$ mutually orthogonal unit normal vector fields ξ_x of M in N such that ξ_{m-n} is a unit normal vector field in the complementary normal line subbundle $V^\perp$ orthogonal to V. Since V is nonparallel, ξ_{m-n} is also nonparallel.

By the umbilicity of M with respect to the normal subbundle V, we have

$$h^u(X, Y) = \alpha^u g(X, Y), \tag{3.1}$$

for some functions α^u and vector fields X, Y in M, where, here and in the sequel, the indices u, v, w run over the range $\{1, 2, \ldots, m - n - 1\}$.

From (3.1) and the equation of Codazzi, we find

$$\{(X\alpha^u) + \alpha^v \theta_v^u(X)\}g(Y, Z) - \{(Y\alpha^u) + \alpha^v \theta_v^u(Y)\}g(X, Z)$$
$$+ k(Y, Z)\theta^u(X) - k(X, Z)\theta^u(Y) = 0, \tag{3.2}$$

$$(\nabla_X k)(Y, Z) - (\nabla_Y k)(X, Z) + h^u(Y, Z)\theta_u(X) - h^u(X, Z)\theta_u(Y) = 0, \tag{3.3}$$

where $k = h^{m-n}$, $\nabla_X^\perp \xi_x = \theta_x^y(X)\xi_y$, $\theta^u = \theta_{m-n}^u = -\theta_u = -\theta_u^{m-n}$.

Hence, if we put

$$D_X \alpha^u = X\alpha^u + \alpha^v \theta_v^u(X) \tag{3.4}$$

and

$$D_{X_j} \alpha^u = \alpha_{/k}^u, \quad X_j = \partial/\partial x^j, \tag{3.5}$$

then we obtain

$$D_X \alpha^u g(Y, Z) - D_Y \alpha^u g(X, Z) + k(Y, Z)\theta^u(X) - k(X, Z)\theta^u(Y) = 0. \tag{3.6}$$

Hence, in exactly the same way as in the proof of (2.10) and (2.11) from (2.2), we may conclude that if $\theta^u \neq 0$, then

$$k = h^{m-n} = \lambda^u g + \mu^u \theta^u \otimes \theta^u, \tag{3.7}$$

where

$$\lambda^u = -\frac{\alpha_{tt}^u \theta^{ut}}{|\theta^u|^2}, \quad \mu^u = \frac{k_t^t - n\lambda^u}{|\theta^u|^2},$$

$$\theta_i^u = \theta^u(X_i), \quad \theta^{ui} = g^{it}\theta_t^u. \tag{3.8}$$

Since, by the assumption, the normal subbundle V is nonparallel, $\theta^u \neq 0$ for some u. Hence, by (3.7) and Proposition V.1.1, we obtain the following

Theorem 3.1 (Chen and Yano, 1973d). *Let M be an n-dimensional $(n > 3)$ submanifold of an m-dimensional Riemannian manifold N of constant curvature c. If M is umbilical with respect to a nonparallel normal $(m - n - 1)$-subbundle, then the submanifold M is conformally flat.*

If M is geodesic with respect to a nonparallel normal $(m - n - 1)$-subbundle, then we have the following

Corollary 3.1 (Chen and Yano, 1972e). *Let M be an n-dimensional submanifold of an m-dimensional Riemannian manifold N of constant curvature c. If M is geodesic with respect to a nonparallel normal $(m - n - 1)$-subbundle, then the submanifold M has constant curvature c.*

This corollary follows immediately from (3.4), (3.5), (3.7), and (3.8).

Lemma 3.1. *Let M be an n-dimensional submanifold of an m-dimensional Riemannian manifold N of constant curvature c. If M is umbilical with respect to a nonparallel normal $(m - n - 1)$-subbundle V and η is a unit normal vector field in the complementary normal line subbundle $V^\perp$ orthogonal to V, then the normal direction η is quasiparallel with respect to a unit normal vector field, say ξ, in V for $n > 2$.*

Proof. Since the normal subbundle V is nonparallel, without loss of generality, we may assume that $\theta^1 \neq 0$. From (3.7), we have

$$h^{m-n} = \lambda^1 g + \mu^1 \theta^1 \otimes \theta^1 \tag{3.9}$$

for some functions λ^1 and μ^1. Thus, if we have $\theta^2 \neq 0$, then, by (3.7), we find

$$(\lambda^1 - \lambda^2)g = -\mu^1 \theta^1 \otimes \theta^1 + \mu^2 \theta^2 \otimes \theta^2. \tag{3.10}$$

Since $n > 2$, (3.10) implies

$$\theta^2 = \gamma^2 \theta^1.$$

Consequently, for any u, we have

$$\theta^u = \gamma^u \theta^1, \tag{3.11}$$

for some functions $\gamma^2, \ldots, \gamma^{m-n-1}$ on M. Hence, from (3.11), we find

$$\nabla^\perp \eta = \nabla^\perp \xi_{m-n} = \theta^u \xi_u = \gamma^u \xi_u \theta^1. \tag{3.12}$$

Thus, if we put

$$\xi = \gamma^u \xi_u / |\gamma^u \xi_u|, \tag{3.13}$$

then we see that the normal direction η is quasiparallel with respect to the normal direction ξ. This proves the lemma. □

In the following, we shall always choose the $m - n$ mutually orthogonal unit normal vector fields ξ_x in such a way that

$$\xi_1 = \xi \quad \text{and} \quad \xi_{m-n} = \eta.$$

Since η is quasiparallel with respect to ξ, we have

$$\theta^1 \neq 0, \quad \theta^2 = \cdots = \theta^{m-n-1} = 0. \tag{3.14}$$

Hence, by (3.1), (3.14), and the equation (II.2.16) of Ricci, we obtain

$$d\theta^1 = 0. \tag{3.15}$$

Thus, if we put

$$\mathcal{D} = \{\text{vector fields } X; \theta^1(X_P) = 0, P \in M\}, \tag{3.16}$$

then, from Theorem I.6.2 and (3.15), we see that the distribution $\mathcal{D}$ is involutive, that is, $\mathcal{D}$ is integrable.

Lemma 3.2. *Let ξ_x be $m - n$ mutually orthogonal unit normal vector fields of M in N chosen as above and $\theta = \theta^1 = \theta^1_{m-n}$ be the 1-form associated with ξ_x. If we put*

$$h^{m-n} = k = \lambda g + \mu \theta \otimes \theta, \tag{3.17}$$

then we have

$$\mu((\nabla_X \theta)(Y)) = \alpha g(X, Y) - (d\mu)(X)\theta(Y) - (d\mu)(Y)\theta(X) + \beta\theta(X)\theta(Y), \tag{3.18}$$

for some functions α and β on M.

Proof. Substituting (3.17) into (3.3), we obtain

$$(X\lambda)g(Y, Z) - (Y\lambda)g(X, Z) + (X\mu)\theta(Y)\theta(Z) - (Y\mu)\theta(X)\theta(Z)$$
$$+ \mu(X\theta(Y) - Y\theta(X))\theta(Z) + \mu_\theta(Y)(X\theta(Z)) - \mu\theta(X)(Y\theta(Z))$$
$$- h^1(Y, Z)\theta(X) + h^1(X, Z)\theta(Y) = 0. \tag{3.19}$$

Hence, if we put

$$\lambda_k = X_k\lambda, \quad \mu_k = X_k\mu, \quad \text{and} \quad \theta_k = \theta(X_k), \tag{3.20}$$

then, by (3.15) and (3.19), we have

$$\lambda_k g_{ji} - \lambda_j g_{ki} + \mu_k \theta_j \theta_i - \mu_j \theta_k \theta_i + \mu\theta_j(\nabla_k\theta_i) - \mu\theta_k(\nabla_j\theta_i)$$
$$- \alpha^1 g_{ji}\theta_k + \alpha^1 g_{ki}\theta_j = 0, \tag{3.21}$$

from which, by transvecting $\theta^k = g^{kt}\theta_t$, we find

$$\lambda_t \theta^t g_{ji} - \lambda_j \theta_i + \mu_t \theta^t \theta_j \theta_i - \mu_j |\theta|^2 \theta_i + \mu\theta_j(\theta^t\nabla_t\theta_i) - \mu|\theta|^2(\nabla_j\theta_i)$$
$$- \alpha^1 |\theta|^2 g_{ji} + \alpha^1 \theta_j \theta_i = 0. \tag{3.22}$$

Equation (3.22) shows that $\mu\nabla_j\theta_i$ is of the following form:

$$\mu\nabla_j\theta_i = pg_{ji} + q_j\theta_i + q_i\theta_j, \tag{3.23}$$

where

$$p = \frac{\lambda_t\theta^t - \alpha^1|\theta|^2}{|\theta|^2},$$

$\nabla_j\theta_i$ being symmetric by (3.15).

Substituting (3.23) into (3.21), we find

$$(\lambda_k - p\theta_k - \alpha^1\theta_k)g_{ji} - (\lambda_j - p\theta_j - \alpha^1\theta_j)g_{ki}$$
$$+ (\mu_k\theta_j - \mu_j\theta_k + q_k\theta_j - q_j\theta_k)\theta_i = 0, \tag{3.24}$$

from which

$$\lambda_k = p\theta_k + \alpha^1\theta_k, \tag{3.25}$$

and we may put

$$\mu_j + q_j = \frac{1}{2}\beta\theta_j. \tag{3.26}$$

By (3.23) and (3.26), we obtain (3.18). This proves the lemma.   □

Theorem 3.2 (Chen and Yano, 1973d). *Let M be an n-dimensional ($n > 2$) submanifold of an m-dimensional space form $R^m(c)$ of curvature c. If M is umbilical with respect to a nonparallel normal ($m - n - 1$)-subbundle V, then M is a locus of ($n - 1$)-spheres in $R^m(c)$.*

Proof. Let θ be the 1-form given in Lemma 3.2 and let

$$\mathcal{D} = \{\text{vector fields } X; \theta(X_P) = 0, P \in M\}.$$

Then $\mathcal{D}$ is integrable. Let M' denote one of the maximal integral hypersurfaces of the foliation $\mathcal{D}$ on M. Then, for any vector field X' in M', we have

$$\theta(X') = 0, \tag{3.27}$$

from which, we find

$$(\nabla_{X'}\theta)(Y') + \theta(\nabla_{X'}Y') = 0, \tag{3.28}$$

for any vector fields X', Y' in M'. Thus, if we denote by h' the second fundamental form of M' in M, then, by (3.27) and (3.28), we find

$$(\nabla_{X'}\theta)(Y') + \theta(h'(X', Y')) = 0. \tag{3.29}$$

Let ζ denote the unit normal vector field of M' in M. Then, by applying Lemma 3.2 to (3.29), we find

$$h'(X', Y') = \gamma g'(X', Y')\zeta \tag{3.30}$$

for some function γ on the open subset $U \cap M'$ of M', where $U = \{P \in M; \mu \neq 0 \text{ at } P\}$ and g' is the induced metric tensor on M'.

From (3.1) and (3.17), we find that the second fundamental form h of M in $R^m(c)$ is given by

$$h = (\alpha^u \xi_u + \lambda \xi_{m-n})g + \mu\theta \otimes \theta \xi_{m-n}. \tag{3.31}$$

Thus, from (3.30) and (3.31), we see that the second fundamental form $\bar{h}$ of M' in $R^m(c)$ is given by

$$\begin{aligned}
\bar{h}(X', Y') &= h'(X', Y') + h(X', Y') \\
&= (\gamma\zeta + \alpha^u\xi_u + \lambda\xi_{m-n})g'(X', Y')
\end{aligned} \tag{3.32}$$

for any vector fields X', Y' in $U \cap M'$. This shows that $U \cap M'$ is totally umbilical in M. Hence, by applying Proposition II.3.2, we see that the closure $\bar{U}$ of U is foliated by $(n-1)$-spheres in $R^m(c)$. On the open subset $M - \bar{U}$, we have $\mu = 0$. Thus, $M - \bar{U}$ is totally umbilical in $R^m(c)$. Hence, by applying Proposition II.3.2, we also see that $M - \bar{U}$ is foliated by $(n-1)$-spheres in $R^m(c)$. Consequently, the submanifold M is a locus of $(n-1)$-spheres in $R^m(c)$. This proves the theorem. $\square$

Remark 3.1. For pseudoumbilical submanifold of codimension 2 in a space form, see also Ōtsuki (1968b), and Yano and Ishihara (1969).

4 Pseudoumbilical Submanifolds with Constant Mean Curvature

From Theorem 2.5, we see that every pseudoumbilical submanifold of codimension 2 in a space form $R^{n+2}(c)$ with constant mean curvature is either a minimal submanifold of $R^{n+2}(c)$ or a minimal hypersurface of a small hypersphere of $R^{n+2}(c)$. In this section, we shall study pseudoumbilical submanifold of codimension 3 in a space form $R^{n+3}(c)$ with constant mean curvature.

Theorem 4.1 (Chen and Yano, 1973e). *Let M be a pseudoumbilical submanifold of codimension 3 in an $(n+3)$-dimensional space form $R^{n+3}(c)$ with flat normal connection. If the mean curvature vector H of M has nonzero constant length and is nonparallel, then the submanifold M is conformally flat for $n > 3$ and it is a locus of $(n-1)$-spheres in $R^{n+3}(c)$ for $n > 2$.*

Proof. Since the mean curvature vector H has nonzero constant length, we may choose three mutually orthogonal unit normal vector fields ξ_x in such a way that

$$H = \alpha\xi_1, \quad \alpha = |H|. \tag{4.1}$$

From the pseudoumbilicity and (4.1), we have

$$h^1 = \alpha g, \tag{4.2}$$

$$h^{(2)} = h^{(3)} = 0, \tag{4.3}$$

where $h^{(x)} = (1/n)h_t^{tx}$.

Substituting (4.2) into equation (II.2.18) of Codazzi, we find

$$\theta_x^1(X)h^x(Y, Z) - \theta_x^1(Y)h^x(X, Z) = 0, \tag{4.4}$$

where $\nabla_X^\perp \xi_x = \theta_x^y(X)\xi_y$, $\theta_x^y = -\theta_y^x$.

Since, by the assumption, the normal connection is flat, we may choose orthonormal vector fields E_i in M such that the E_i are the principal directions of M in $R^{n+3}(c)$ with respect to ξ_2 and ξ_3 with principal curvatures λ_i and μ_i, respectively. From (4.4) we find

$$\theta_2^1(E_k)\lambda_j + \theta_3^1(E_k)\mu_j = 0, \quad \text{for } k \neq j. \tag{4.5}$$

Combining (4.3) and (4.5), we find

$$\theta_2^1(E_k)\lambda_j + \theta_3^1(E_k)\mu_j = 0, \quad \text{for all } k \text{ and } j. \tag{4.6}$$

By the assumption, the mean curvature vector H is nonparallel, and without loss of generality, we may assume that $\theta_2^1(E_1) \neq 0$. From (4.6) we find

$$\lambda_j = \gamma\mu_j, \quad \gamma = -\frac{\theta_3^1(E_1)}{\theta_2^1(E_1)}. \tag{4.7}$$

This implies that

$$A_{\xi_2} = \gamma A_{\xi_3}. \tag{4.8}$$

We put $\gamma = -\tan \varphi$ and

$$\xi_2' = (\cos \varphi)\xi_2 + (\sin \varphi)\xi_3,$$
$$\xi_3' = -(\sin \varphi)\xi_2 + (\cos \varphi)\xi_3.$$

Then, by a direct computation, we find that $A_{\xi_2'} = 0$ identically. Hence, from now on, we may assume that $A_2 = 0$, that is,

$$h^2 = 0. \tag{4.9}$$

By substituting (4.9) into the equation (II.2.18) of Codazzi, we find

$$\theta_3^1(E_k)\mu_j = 0. \tag{4.10}$$

This shows that we have either $\theta_3^1 = 0$ or $h^3 = 0$. Now, suppose that the subset $W = \{P \in M; h^3 = 0 \text{ at } P\}$ has nonempty interior, that is, $\text{int}(W) \neq \emptyset$. Then, since $\text{int}(W)$ is totally umbilical in $R^{n+3}(c)$; each component of $\text{int}(W)$ is contained in an n-sphere of $R^{n+3}(c)$. In particular, the mean curvature vector H is parallel on $\text{int}(W)$. This is a contradiction. Hence, we see that $\text{int}(W)$ is empty. Thus, by the continuity of θ_3^1, we get $\theta_3^1 = 0$. Consequently, we have

$$\theta_2^1 \neq 0, \quad \theta_3^1 = 0. \tag{4.11}$$

On the other hand, the normal connection $\nabla^\perp$ is flat, that is,

$$\nabla_X^\perp \nabla_Y^\perp \xi_x - \nabla_Y^\perp \nabla_X^\perp \xi_x - \nabla_{[X,Y]}^\perp \xi_x = 0. \tag{4.12}$$

From (4.11) and (4.12), we find

$$\theta_3^2 \wedge \theta_2^1 = 0. \tag{4.13}$$

Since $\theta_2^1 \neq 0$, this implies that

$$\theta_3^2 = \nu \theta_2^1 \tag{4.14}$$

for some function ν on M.

If $\nu = 0$ at a point $P \in M$, then, by (4.2), (4.9), and the equation (II.2.18) of Codazzi, we find

$$\alpha\{g(Y, Z)\theta_1^2(X) - g(X, Z)\theta_1^2(Y)\} = 0 \tag{4.15}$$

at the point P. From this we find

$$\theta_1^2 = 0 \tag{4.16}$$

at the point P. This is a contradiction. Thus, the function ν is nowhere zero. Consequently, the normal 2-subbundle V generated by H and ξ_2 is nonparallel in the normal bundle. Since the submanifold M is umbilical with respect to V, Theorems 3.1 and 3.2 imply that M is conformally flat for $n > 3$ and M is a locus of $(n-1)$-spheres in $R^{n+3}(c)$ for $n > 2$. This completes the proof of the theorem. $\qquad\square$

5 Characterizations of Umbilical Submanifolds

Let M be an n-dimensional submanifold of a euclidean m-space E^m. In this section, we shall give some characterizations for M to be umbilical with respect to a normal direction.

In the following, we assume that ξ is a unit normal vector field on the submanifold M in E^m which is parallel in the normal bundle. Let E_i be the principal directions with respect to the normal direction ξ with principal curvatures k_i and let ξ_x be $m - n$ mutually orthogonal unit normal vector fields of M such that

$$\xi_1 = \xi. \tag{5.1}$$

If we denote by $\tilde{X}$ the position vector field on E^m with respect to the origin and by $\tilde{\nabla}$ the usual (euclidean) connection on E^m, then, by the flatness of E^m, we have

$$\tilde{\nabla}^2 \tilde{X} = \tilde{\nabla}\tilde{\nabla}\tilde{X} = 0, \quad \tilde{\nabla}^2 \xi_x = \tilde{\nabla}\tilde{\nabla}\xi_x = 0. \tag{5.2}$$

In the following, we also denote by $\tilde{X}$ the position vector field of M in E^m with respect to the origin. Let ω^i denote the dual vector fields of E_i, that is, ω^i are 1-forms in M such that $\omega^i(E_j) = \delta^i_j$. Then we have

$$\tilde{\nabla}\tilde{X} = \omega^i E_i. \tag{5.3}$$

The volume element of M is then given by

$$dV = \omega^1 \wedge \cdots \wedge \omega^n. \tag{5.4}$$

In the following, we put

$$\begin{aligned}
\tilde{\nabla} E_i &= \omega^j_i E_j + \omega^{x+n}_j \xi_x, \\
\tilde{\nabla} E_x &= \omega^j_{x+n} E_j + \theta^y_x \xi_y.
\end{aligned} \tag{5.5}$$

Then we have

$$\omega^j_i = -\omega^i_j, \quad \omega^{x+n}_i = -\omega^i_{x+n}, \quad \theta^y_x = -\theta^x_y, \tag{5.6}$$

$$\omega^j_{x+n} E_j = -A_x. \tag{5.7}$$

Since, by the assumption, $\xi = \xi_1$ is parallel in the normal bundle with principal curvatures k_i, we have

$$\tilde{\nabla}_{E_j} \xi_1 = -k_j E_j. \tag{5.8}$$

In the following, we decompose the position vector field $\tilde{X}$ of the submanifold M into two components:

$$\tilde{X} = \tilde{X}_t + \tilde{X}_n, \tag{5.9}$$

where $\tilde{X}_t$ is tangent to M and $\tilde{X}_n$ is normal to M. Let ξ' be a unit normal vector field of M orthogonal to ξ and in the direction $\tilde{X}_n - \tilde{g}(\tilde{X}_n, \xi)\xi$, that is,

$$\tilde{X}_n = \tilde{g}(\tilde{X}_n, \xi)\xi + \tilde{g}(\tilde{X}_n, \xi')\xi', \tag{5.10}$$

where $\tilde{g}$ is the euclidean metric tensor on E^m.

In the following, we shall always assume that ξ_x are $m - n$ mutually orthogonal unit normal vector fields of M such that

$$\xi_1 = \xi, \quad \xi_2 = \xi', \quad \det(E_1, \ldots, E_n, \xi_1, \ldots, \xi_{m-n}) = 1. \tag{5.11}$$

We define n functions $F_i(\xi)$ by

$$F_i(\xi) = \frac{(n-1)!}{n!} \tilde{g}(\tilde{X}, \xi') \sum k_{j_1} \cdots k_{j_{i-1}} \tilde{g}(A_{\xi'}(E_{j_i}), E_{j_i}), \tag{5.12}$$

where the summation is taken over all distinct $j_1, \ldots, j_i$.

For convenience, we shall introduce, according to H. Hopf and K. Voss, the combined operation of exterior product and vector product. We denote such a combined operation by $[\, , \ldots, \,]$ $(m - 1$ terms). For simple cases we have

$$[E_1, \ldots, \hat{E}_i, \ldots, E_n, \xi_1, \ldots, \xi_{m-n}] = (-1)^{m+i} E_i,$$

$$[E_1, \ldots, E_n, \xi_1, \ldots, \hat{\xi}_y, \ldots, \xi_{m-n}] = (-1)^{m+n+y} \xi_y,$$

$$g(Z_1, [Z_2, \ldots, Z_m]) = (-1)^{m-1} \det(Z_1, \ldots, Z_m),$$

$$\left[\underbrace{\tilde{\nabla}\tilde{X}, \ldots, \tilde{\nabla}\tilde{X}}_{n-i \text{ times}}, \underbrace{\tilde{\nabla}\xi_1, \ldots, \tilde{\nabla}\xi_1}_{i \text{ times}}, \xi_2, \ldots, \xi_{m-n} \right] = n!(-1)^{m+n+1} M_i(\xi)\xi dV,$$

where $\wedge$ denotes the omitted term, $Z_1, \ldots, Z_m$ are m vectors in E^m at a point $P \in E^m$, and $M_i(\xi)$ is the i-th mean curvature with respect to ξ (see Problem II.2).

Theorem 5.1 (Chen and Yano, 1971a). *Let M be an n-dimensional oriented closed submanifold of a euclidean m-space E^m. If ξ is a unit parallel normal vector field of M in E^m, then we have*

$$\int_M \{M_{i-1}(\xi) + s(\xi)M_i(\xi)\}dV = -\int_M F_i(\xi)dV, \quad i = 1, \ldots, n, \tag{5.13}$$

where $s(\xi) = \tilde{g}(X, \xi)$ is the support function of M with respect to the normal direction ξ.

Proof. Under the hypothesis, we choose ξ_x as $m - n$ mutually orthogonal unit normal vector fields of M satisfying (5.11). Then, by using the same notations, we have

$$d\left(\tilde{g}\left(\tilde{X}, \left[\underbrace{\tilde{\nabla}\tilde{X}, \ldots, \tilde{\nabla}\tilde{X}}_{n-i \text{ times}}, \underbrace{\tilde{\nabla}\xi, \ldots, \tilde{\nabla}\xi}_{i-1 \text{ times}}, \xi_1, \ldots, \xi_{m-n}\right]\right)\right)$$

$$= \tilde{g}\left(\tilde{\nabla}\tilde{X}, \left[\underbrace{\tilde{\nabla}\tilde{X}, \ldots, \tilde{\nabla}\tilde{X}}_{n-i \text{ times}}, \underbrace{\tilde{\nabla}\xi, \ldots, \tilde{\nabla}\xi}_{i-1 \text{ times}}, \xi_1, \ldots, \xi_{m-n}\right]\right)$$

$$+ +(-1)^{n-1}\tilde{g}\left(\tilde{X}, \left[\underbrace{\tilde{\nabla}\tilde{X}, \ldots, \tilde{\nabla}\tilde{X}}_{n-i \text{ times}}, \underbrace{\tilde{\nabla}\xi, \ldots, \tilde{\nabla}\xi}_{i \text{ times}}, \xi_2, \ldots, \xi_{m-n}\right]\right)$$

$$+ (-1)^{n-1}\sum_{y=2}^{m-n}\tilde{g}\left(\tilde{X}, \left[\underbrace{\tilde{\nabla}\tilde{X}, \ldots, \tilde{\nabla}\tilde{X}}_{n-i \text{ times}}, \underbrace{\tilde{\nabla}\xi, \ldots, \tilde{\nabla}\xi}_{i-1 \text{ times}}, \xi_1, \ldots, \xi_{y-1}, \right.\right.$$
$$\left.\left. \tilde{\nabla}\xi_y, \xi_{y+1}, \ldots, \xi_{m-n}\right]\right)$$

$$= (-1)^{m+i}\{M_{i-1}(\xi) + s(\xi)M_i(\xi)\}n!dV$$

$$+ (-1)^{n-1}\tilde{g}\left(\tilde{X}, \left[\underbrace{\tilde{\nabla}\tilde{X}, \ldots, \tilde{\nabla}\tilde{X}}_{n-i \text{ times}}, \underbrace{\tilde{\nabla}\xi, \ldots, \tilde{\nabla}\xi}_{i-1 \text{ times}}, \xi_1, \tilde{\nabla}\xi', \xi_3, \ldots, \xi_{m-n}\right]\right)$$

$$\text{(5.14)}$$

and

$$\left[\underbrace{\tilde{\nabla}\tilde{X}, \ldots, \tilde{\nabla}\tilde{X}}_{n-i \text{ times}}, \underbrace{\tilde{\nabla}\xi, \ldots, \tilde{\nabla}\xi}_{i-1 \text{ times}}, \xi_1, \tilde{\nabla}\xi', \xi_3 \ldots, \xi_{m-n}\right]$$

$$= \sum \omega^{j_1} \wedge \cdots \wedge \omega^{j_{n-i}} \wedge \omega_{n+1}^{j_{n-i+1}} \wedge \cdots \wedge \omega_{n+1}^{j_{n-1}} \wedge \omega_{n+2}^{j_n}$$
$$\cdot [E_{j_1}, \ldots, E_{j_{n-i}}, \ldots, E_{j_{n-1}}, \xi_1, E_{j_n}, \xi_3, \ldots, \xi_{m-n}]$$

$$= (-1)^{i-1}\sum k_{j_1}\cdots k_{j_{i-1}} g(A_{\xi'}(E_{j_i}), E_{j_i})\omega^{j_1} \wedge \cdots \wedge \omega^{j_n}$$
$$\cdot [E_{j_1}, \ldots, E_{j_n}, \xi_1, \xi_3, \ldots, \xi_{m-n}]$$

$$= (n-i)!(-1)^{m+n+i+1}\left(\sum k_{j_1}\cdots k_{j_{i-1}} g(A_{\xi'}(E_{j_i}), E_{j_i})\xi'dV\right). \quad \text{(5.15)}$$

From (5.12), (5.14), and (5.15), we find

$$d\left(\tilde{g}\left(\tilde{X}, \left[\underbrace{\tilde{\nabla}\tilde{X}, \ldots, \tilde{\nabla}\tilde{X}}_{n-i \text{ times}}, \underbrace{\tilde{\nabla}\xi, \ldots, \tilde{\nabla}\xi}_{i-1 \text{ times}}, \xi_1, \ldots, \xi_{m-n}\right]\right)\right)$$

$$= (-1)^{m+i}n!\{M_{i-1}(\xi) + s(\xi)M_i(\xi) + F_i(\xi)\}dV, \quad i = 1, \ldots, n.$$

From this we see that $F_i(\xi)$ are well-defined functions on M. By integrating both sides of the above equation and applying Green's theorem, we obtain the proposition.   □

From Theorem 5.1, we have immediately the following:

Theorem 5.2 (Chen and Yano, 1971a). *Let M be an n-dimensional oriented closed submanifold of a euclidean m-space E^m. If ξ is a unit parallel normal vector field of M with $F_i(\xi) = 0$ for some i, then we have*

$$\int_M M_{i-1}(\xi)dV + \int_M s(\xi)M_i(\xi)dV = 0. \tag{5.16}$$

In the following, we shall give some examples in which the functions $F_i(\xi)$ vanish for some normal direction ξ.

Example 5.1. If the codimension is one and ξ is a unit normal vector field of M in E^{n+1}, then $F_i(\xi) = 0$ for all i. In this case, formulas (5.16) are called the Minkowski-Hsiung formulas (Hsiung, 1954).

Example 5.2. If M is contained in a small hypersphere of E^m centered at a point C, then the radius vector field $\tilde{X} - C$ is normal to M. Let $\xi = (\tilde{X} - C)/|\tilde{X} - C|$. Then we have $s(\xi') = 0$. Thus we have $F_i(\xi) = 0$ for all i.

Example 5.3. If ξ is in the direction of $\tilde{X}_n$, then we have $F_i(\xi) = 0$ for all i.

Example 5.4. If ξ is a unit normal vector field of M and if the submanifold M is geodesic with respect to ξ', then we have $A_{\xi'} = 0$. Hence, $F_i(\xi) = 0$ for all i.

In the remaining part of this section, we shall use formulas (5.16) to obtain some characterizations for umbilical submanifolds in E^m.

Corollary 5.1 (Chen and Yano, 1971a). *Let M be an n-dimensional oriented closed submanifold of a euclidean m-space E^m. If there exist a unit parallel normal vector field ξ of M and an integer i, $2 \leqq i \leqq n$, such that*

(i) $M_i(\xi) > 0$,

(ii) $s(\xi) \leqq -M_{i-1}(\xi)/M_i(\xi)$, [or $s(\xi) \geqq -M_{i-1}(\xi)/M_i(\xi)$],

(iii) $F_i(\xi) = F_{i-1}(\xi) = 0$,

then M is umbilical with respect to the normal direction ξ.

Proof. Under the hypothesis, we obtain from Theorem 5.2 that

$$s(\xi) = -M_{i-1}(\xi)/M_i(\xi),$$

$$\int_M \{M_{i-2}(\xi) + s(\xi)M_{i-1}(\xi)\}dV = 0,$$

from which, we find

$$\int_M \left(\frac{1}{M_i(\xi)}\right)\{M_{i-1}(\xi)^2 - M_{i-2}(\xi)M_i(\xi)\}dV = 0.$$

Thus, by (i) and Problem II.2, we find that M is umbilical with respect to the normal direction ξ. This proves the Corollary.   □

Corollary 5.2 (Chen and Yano, 1971a). *Let M be an n-dimensional oriented closed submanifold of a euclidean m-space E^m. If there exist a unit parallel normal vector field ξ of M and an integer i, $1 < i < n$, such that*

    (i) $M_{i+1}(\xi) > 0$,

    (ii) $s(\xi) \geqq -M_{i-1}(\xi)/M_i(\xi)$,

    (iii) $F_{i+1}(\xi) = 0$,

then M is umbilical with respect to the normal direction ξ.

Proof. Under the hypothesis, we obtain from Problem II.2 that

$$s(\xi) \geqq -M_{i-1}(\xi)/M_i(\xi) \geqq -M_i(\xi)/M_{i+1}(\xi),$$

which, together with

$$\int_M \{M_i(\xi) + s(\xi)M_{i+1}(\xi)\}dV = 0,$$

implies that

$$s(\xi) \geqq -M_{i-1}(\xi)/M_i(\xi) \geqq -M_i(\xi)/M_{i+1}(\xi) = s(\xi).$$

Therefore, we get $(M_i(\xi))^2 = M_{i-1}(\xi)M_{i+1}(\xi)$. Hence, by Problem II.2, we see that M is umbilical with respect to the normal direction ξ. This proves the Corollary.   □

Corollary 5.3 (Chen and Yano, 1971a). *Let M be an n-dimensional oriented closed submanifold of a euclidean m-space E^m. If there exist a unit parallel normal vector field ξ of M and two integers s and i, $1 \leqq i < s \leqq n$, such that*

    (i) $M_s(\xi), \ldots, M_i(\xi) > 0$,

    (ii) $M_s(\xi) = \sum_{j=1}^{s-1} c_j M_j(\xi)$, *for some constants c_j, $1 \leqq j \leqq s - 1$,*

    (iii) $F_j(\xi) = 0$, $j = 1, \ldots, s - 1$,

then M is umbilical with respect to the normal direction ξ.

Proof. By (i) and Problem II.2, we have

$$\frac{M_j(\xi)}{M_s(\xi)} - \frac{M_{j-1}(\xi)}{M_{s-1}(\xi)} = \left(\frac{M_j(\xi)}{M_{s-1}(\xi)}\right)\left\{\frac{M_{s-1}(\xi)}{M_s(\xi)} - \frac{M_{j-1}(\xi)}{M_j(\xi)}\right\} \geqq 0,$$

for $i \leqq j \leqq s$, so that

$$1 = \sum c_j M_j(\xi)/M_s(\xi) \geqq \sum c_j M_{j-1}(\xi)/M_{s-1}(\xi),$$

or

$$M_{s-1}(\xi) - \sum c_j M_{j-1}(\xi) \geqq 0,$$

where the equality sign holds only if M is umbilical with respect to ξ. Thus, by using (ii) and Theorem 5.2, we find

$$\int_M \left\{ M_{s-1}(\xi) - \sum_{j=1}^{s-1} c_j M_{j-1}(\xi) \right\} dV$$

$$= -\int_M s(\xi) \left\{ M_s(\xi) - \sum_{j=i}^{s-1} c_j M_j(\xi) \right\} dV = 0. \qquad (5.17)$$

Therefore, we have $M_{s-1}(\xi) = \sum c_j M_{j-1}(\xi)$. This implies that M is umbilical with respect to the normal direction ξ. This proves the Corollary. □

Corollary 5.4 (Chen and Yano, 1971a). *Let M be an n-dimensional oriented closed submanifold of a euclidean m-space E^m. If there exist a unit parallel normal vector field ξ and two integers s and i, $0 \leqq i < s \leqq n$, such that*

(i) $M_{s+1}(\xi), \ldots, M_{i+1}(\xi) > 0$,

(ii) $M_s(\xi) = \sum_{j=1}^{s-1} c_j M_j(\xi)$, for some constants c_j, $1 \leqq j \leqq s - 1$,

(iii) $s(\xi) > 0$ or $s(\xi) < 0$,

(iv) $F_{s-1}(\xi) = \cdots = F_i(\xi) = 0$,

then M is umbilical with respect to the normal direction ξ.

This corollary follows immediately from the assumption and formulas (5.17).

Corollary 5.5 (Chen and Yano, 1971a). *Let M be an n-dimensional oriented closed submanifold of a euclidean m-space E^m. If there exist a unit parallel normal vector field ξ of M and an integer i, $0 < i \leqq n$, such that*

(i) $M_i(\xi) > 0$,

(ii) $M_i(\xi) = cM_{i-1}(\xi)$, for some constants c,

(iii) $F_{i-1}(\xi) = F_i(\xi) = 0$,

then M is umbilical with respect to the normal direction ξ.

Proof. Since $M_i(\xi) > 0$, c cannot be zero and $M_{i-1}(\xi)$ must be a fixed sign. By (ii) and Problem II.2, we find

$$M_{i-1}(\xi)\{M_{i-1}(\xi) - cM_{i-2}(\xi)\} = (M_{i-1}(\xi))^2 - M_i(\xi)M_{i-2}(\xi) \geqq 0,$$

from which, together with (ii) and Theorem 5.2, we find

$$\int_M \{M_{i-1}(\xi) - cM_{i-2}(\xi)\}dV = \int_M \{cM_{i-1}(\xi) - M_i(\xi)\}s(\xi)dV = 0.$$

This implies that $M_{s-1}(\xi) = cM_{s-2}(\xi)$. Thus, by applying Corollary 5.3, we see that M is umbilical with respect to the normal direction ξ. This proves the corollary.   □

Corollary 5.6 (Chen and Yano, 1971a). *Let M be an n-dimensional oriented closed submanifold of a euclidean m-space E^m. If there exists a unit parallel normal vector field ξ of M such that*

(i) $M_n(\xi) > 0$,

(ii) *the sum of the principal radii of curvature is constant at ξ, that is, $\sum_{i=1}^{n} k_i^{-1} = constant$, where k_i are the principal curvatures with respect to ξ,*

(iii) $F_n(\xi) = F_{n-1}(\xi) = 0$,

then M is umbilical with respect to the normal direction ξ.

Proof. Since $\sum_{i=1}^{n} \frac{1}{k_i} = nM_{n-1}(\xi)/M_n(\xi) = $ constant, Corollary 5.5 implies that M is umbilical with respect to the normal direction ξ. This proves the corollary.   □

6   The Gauss Map

Let M be an n-dimensional submanifold of an m-dimensional Riemannian manifold N and ζ be a normal vector field of M. Let ξ_x be $m - n$ mutually orthogonal unit normal vector fields of M such that

$$\zeta = |\zeta|\xi_1. \tag{6.1}$$

We define a normal vector field $a(\zeta)$ by

$$a(\zeta) = \frac{|\zeta|}{n} \sum_{x=2}^{m-n} \{\text{trace}\,(A_1 A_x)\}\xi_x. \tag{6.2}$$

Then $a(\zeta)$ is a well-defined normal vector field on M and it is continuous on M. We call $a(\zeta)$ the *allied vector field* of ζ. For example, if M is contained in a hypersphere S^{m-1} of a euclidean m-space E^m and ζ is the unit outer hypersphere normal vector field of S^{m-1}, then the allied vector field $a(\zeta)$ of ζ is just the mean curvature vector of M in S^{m-1}.

The allied vector field of the mean curvature vector H is a well-defined normal vector field orthogonal to H. We call it the *allied mean curvature vector* of M in N. If the allied mean curvature vector $a(H)$ vanishes identically, then the submanifold M is called an $\mathfrak{U}$-*submanifold* of N. It is clear that minimal submanifolds, pseudoumbilical submanifolds, and hypersurfaces are $\mathfrak{U}$-submanifolds. There are $\mathfrak{U}$-submanifolds which are not one of the submanifolds we just mentioned.

Let M be an n-dimensional submanifold of a euclidean m-space E^m and η be a unit normal vector field of M in E^n. Then, by using the parallel translation in E^m, we obtain a map from the set $\{\eta_P: P \in M\}$ into the unit hypersphere of E^m centered at a fixed point. We call this map the *Gauss map* of the normal vector field η.

For an $\mathfrak{U}$-submanifold M of a small hypersphere S^{m-1} in a euclidean m-space E^m. It can be proved that M is a pseudoumbilical submanifold of S^{m-1} if and only if M is an $\mathfrak{U}$-submanifold of E^m.

In the following, we prove

Proposition 6.1. *Let M be an n-dimensional closed $\mathfrak{U}$-submanifold of a small hypersphere S^{m-1} of a euclidean m-space E^m. If the mean curvature vector H' of M in S^{m-1} is nonzero and parallel, then the submanifold M is pseudoumbilical in S^{m-1} (and hence, pseudoumbilical in E^m) if and only if the Gauss map of $H'/|H'|$ lies in an open hemisphere of S^{m-1}.*

Proof. Without loss of generality, we may assume that S^{m-1} is the unit hypersphere of E^m centered at the origin. Let $\tilde{X}$ denote the position vector field of M in E^m centered at the origin. Then $\tilde{X}$ is a unit normal vector field of M in E^m. Since the mean curvature vector H' of M in S^{m-1} is parallel in the normal bundle of M in S^{m-1}, H' has constant length. In the following, we choose $m - n - 1$ mutually orthogonal unit normal vector fields $\xi_1, \ldots, \xi_{m-n-1}$ of M in S^{m-1} in such a way that

$$\xi_1 = H'/|H'|. \tag{6.3}$$

Then if we put

$$\xi_{m-n} = \tilde{X}, \tag{6.4}$$

$\xi_1, \ldots, \xi_{m-n}$ may be regarded as $m - n$ mutually orthogonal unit normal vector fields of M in E^m. Let h and h' denote the second fundamental forms of M in E^m and S^{m-1}, respectively. Then, by Eq. (III.3.4), we have

$$h(X, Y) = h'(X, Y) - g(X, Y)\tilde{X}, \tag{6.5}$$

for any vector fields X, Y in M. From (6.5), we see that the second fundamental tensors A_ξ and A'_ξ of M in E^m and S^{m-1} are equal for any normal vector field ξ of M in S^{m-1}. Thus, in the following, we may simply denote A'_ξ by A_ξ.

By the assumption, the Gauss map of ξ_1 lies in an open hemisphere of S^{m-1}. Hence, there exists a fixed vector c in E^m such that

$$\tilde{g}(\xi_1, c) > 0, \tag{6.6}$$

where $\tilde{g}$ is the usual euclidean metric tensor of E^m.

In the following, let $\tilde{\nabla}$ denote the euclidean connection of E^m induced from $\tilde{g}$ and ∇ the induced connection on M with the induced Riemannian metric tensor g. Then, by the parallelism of ξ_1 in the normal bundle of M in S^{m-1}, we see that ξ_1 is also parallel in the normal bundle of M in E^m. Hence, by the equation of Codazzi, we find

$$
\begin{aligned}
\Delta\tilde{g}(\xi_1, c) &= \nabla^i \nabla_i g(\xi_1, c) = g^{ji} \tilde{\nabla}_{X_j} \tilde{\nabla}_{X_i} \tilde{g}(\xi_1, c) \\
&= -g^{ji} \tilde{\nabla}_{X_j} \tilde{g}(A_1(X_i), c) \\
&= -g^{ji} \tilde{g}(\nabla_{X_j}(A_1(X_i)), c) - g^{ji} \tilde{g}(h(X_j, A_1(X_i)), c) \\
&= -\tilde{g}(g^{ji}\{X_j(\operatorname{trace} A_1)\}X_i, c) - g^{ji} \tilde{g}(h(X_j, A_1(X_i)), c), \quad (6.7)
\end{aligned}
$$

where the last equality is obtained from the parallelism of ξ_1 and the equation of Codazzi for M in E^m. Since

$$
\operatorname{trace} A_1 = |H'| = \text{constant}, \tag{6.8}
$$

by (6.5) and (6.7), we find

$$
\Delta\tilde{g}(\xi_1, c) = \tilde{g}((\operatorname{trace} A_1)\tilde{X} - (\operatorname{trace} A_1^2)\xi_1 - na(\xi_1), c), \tag{6.9}
$$

where $a(\xi_1)$ is the allied vector field of ξ_1 in S^{m-1}. Since, by the assumption, M is an $\mathfrak{U}$-submanifold of S^{m-1}, $a(\xi_1)$ vanishes. Thus, by (6.9), we have

$$
\Delta\tilde{g}(\xi_1, c) = \tilde{g}((\operatorname{trace} A_1)\tilde{X} - (\operatorname{trace} A_1^2)\xi_1, c). \tag{6.10}
$$

On the other hand, from Eq. (II.5.7), we have

$$
\Delta\tilde{g}(\tilde{X}, c) = \tilde{g}((\operatorname{trace} A_1)\xi_1 - n\tilde{X}, c). \tag{6.11}
$$

Hence, by combining (6.10) and (6.11), we find

$$
\begin{aligned}
\Delta\tilde{g}(n\xi_1 + (\operatorname{trace} A_1)\xi_1, c) &= \{-n \operatorname{trace} A_1^2 - (\operatorname{trace} A_1)^2\}\tilde{g}(\xi_1, c) \\
&= -\sum_{i<j}\{\lambda_i - \lambda_j\}^2\tilde{g}(\xi_1, c), \tag{6.12}
\end{aligned}
$$

where λ_i are the principal curvatures of the normal direction ξ_1. Hence, by (6.6) and (6.12), we see that

$$
\Delta\tilde{g}(n\xi_1 + (\operatorname{trace} A_1)\xi_1, c) \geq 0.
$$

Thus, by applying the Hopf lemma, we see that $\lambda_1 = \cdots = \lambda_n$. This implies that the submanifold M is pseudoumbilical in S^{m-1}. The converse of this is trivial. □

Similarly, in exactly the same way as in the proof of this theorem, we can also prove the following:

Proposition 6.2. *Let M be an n-dimensional closed minimal submanifold of a small hypersphere S^{m-1} of a euclidean m-space E^m. If there exists a unit parallel normal vector field ξ of M in S^{m-1} such that the allied vector field $a(\xi) = 0$ and* *the Gauss map of ξ lies in an open hemisphere, then M is contained in a great hypersphere of S^{m-1}. The converse of this is also true.*

Remark 6.1. If the codimension of M in S^{m-1} is one, then Proposition 6.1 was proved by Nomizu and Smyth (1969b), and Proposition 6.2 was proved by de Giorgi (1965) and Simons (1968).

Problems

1. Let M be a 3-dimensional pseudoumbilical submanifold of a 6-dimensional Riemannian manifold N of constant curvature c. Prove that if the mean curvature vector of M in N has nonzero constant length and is nonparallel and if the normal connection of M in N is flat, then M is conformally flat (Chen and Yano, 1973e).

2. Let M be a pseudoumbilical surface with constant Gaussian curvature in an m-dimensional space form $R^m(c)$ of curvature c. Prove that if the mean curvature of M is a nonzero constant and the normal connection is flat, then M is either flat or totally umbilical in $R^m(c)$.

3. Let M be an $\mathfrak{U}$-submanifold of a euclidean m-space E^m. Prove that if M is contained in a small hypersphere of S^{m-1} of E^m, then M is an $\mathfrak{U}$-submanifold of S^{m-1} if and only if M is a pseudoumbilical submanifold of S^{m-1}.

4. Let M be an n-dimensional closed oriented hypersurface of a small hypersphere S^{n+1} of a euclidean $(n+2)$-space E^{n+2} with constant mean curvature. Prove that if the Gauss map of the unit hypersurface normal vector field of M in S^{n+1} lies in a closed hemisphere, then M is a hypersphere of S^{n+1} (Nomizu and Smyth, 1969b).

5. Let M be an n-dimensional submanifold of a small hypersphere S^{m-1} of a euclidean m-space E^m and let H denote the mean curvature vector of M in E^m, then H is nowhere zero. Let ξ be a unit normal vector field of M in E^m given by $\xi = H/|H|$ and ξ' be the unit normal vector field defined by (5.10). Prove that M is a pseudoumbilical submanifold of S^{m-1} if and only if the position vector field $\tilde{X}$ with respect to the center of S^{m-1} satisfies $\tilde{g}(\tilde{X}, \xi')A_{\xi'} = 0$ (Yano and Chen, 1971a).

6. Let M be a hypersurface of a Riemannian manifold N. Prove that if the mean curvature vector H of M in N is nowhere zero and the length of the second fundamental form of M in N is constant, then the product submanifold $M \times M$ in $N \times N$ is an $\mathfrak{U}$-submanifold.

7. Use Problem 6 to construct an $\mathfrak{U}$-submanifold which is neither a minimal submanifold, a pseudoumbilical submanifold, or a hypersurface.

Chapter 7

Geometric Inequalities

1 Some Results in Morse's Theory

Let M be an n-dimensional closed manifold. Given a differentiable function f on M, a point $P \in M$ is called a *critical point* of f if $(df)_P = 0$. If we choose a local coordinate system $\{x^i\}$ in a neighborhood of the point P, this means that $(\partial_i f)_P = 0$, where $\partial_i = \partial/\partial x^i$.

If P is a critical point of the function f, then the matrix

$$(\partial_j \partial_i f)_P$$

represents a symmetric bilinear function f_{**} on the tangent space $T_P(M)$ of M at P. We call this symmetric function f_{**} the *Hessian form* of f at the point P.

A critical point $P \in M$ of the function f is called a *nondegenerate point* of f if the Hessian form of f at the point P is nondegenerate. In this case the dimension of a maximal-dimensional subspace of $T_P(M)$ on which the Hessian form is negative definitive is called the *index* of the critical point P. A function f on M is said to be nondegenerate if all of its critical points are nondegenerate. A nondegenerate function on M is sometimes called a *Morse function* on M. We introduce the following notations.

$\Phi(M)$ = the set of nondegenerate functions on M,

$\beta_k(f)$ = the number of the critical points of index k of f,

$$\beta(f) = \sum_{k=0}^{n} \beta_k(f) = \text{the number of the critical points of } f.$$

For any field $\mathcal{F}$ we denote by $H_k(M, \mathcal{F})$ the k-th homology group of M with coefficients in the field $\mathcal{F}$. We put

$$b_k(M, \mathcal{F}) = \dim_{\mathcal{F}} H_k(M, \mathcal{F}),$$

$$b(M, \mathcal{F}) = \sum_{k=0}^{n} b_k(M, \mathcal{F}),$$

$$b(M) = \max\{b(M, \mathcal{F}); \mathcal{F} \text{ field}\}.$$

Theorem 1.1 (Weak Morse Inequalities). *Let M be an n-dimensional closed manifold. Then for any field $\mathcal{F}$ and any function $f \in \Phi(M)$ we have*

$$\beta_k(f) \geqq b_k(M, \mathcal{F}), \tag{1.1}$$

$$\sum_{k=0}^{n}(-1)^k \beta_k(f) = \sum_{k=0}^{n}(-1)^k b_k(M, \mathcal{F}) = \chi(M), \tag{1.2}$$

where $\chi(M)$ is the Euler characteristic of M.

For the proof of this theorem see, for instance, Milnor (1963).

Theorem 1.2 (Reeb, 1952). *Let M be a closed n-dimensional manifold. If there exists a (differentiable) function f on M with only two nondegenerate critical points, then M is homeomorphic to an n-sphere.*

For the proof of this theorem see Reeb (1952) or Milnor (1963).

Let M and N be two n-dimensional manifolds of differentiable class C^1. For a differentiable mapping x of class C^1 of M into N, if the differential $(x_*)_P$ of x at P is not injective, then the point $P \in M$ is called a *critical point* of x.

Theorem 1.3 (Sard, 1942). *Let M and N be two n-dimensional manifolds of differentiable class C^1 and x a differentiable mapping of class C^1 of M into N. Then the image $x(E)$ of the set of critical points of x is a set of measure zero in N.*

For the proof of this theorem see, for instance, Milnor (1965).

2 Some Results of Chern and Lashof

Let $x: M \to E^m$ be an immersion of an n-dimensional closed manifold M into a euclidean m-space E^m. The normal bundle $T^{\perp}(M)$ of M in E^m is an $(m-n)$-plane bundle over M whose bundle space is the subset of $M \times E^m$, consisting of all points (P, ξ), such that $P \in M$ and ξ is a normal vector of M at P. With respect to the induced metric from E^m the normal bundle $T^{\perp}(M)$ is a Riemannian $(m-n)$-plane bundle over M. If B is the subbundle of the normal bundle whose bundle space consists of all points (P, ξ) such that $P \in M$ and ξ is a unit normal vector of M at P, then B is a bundle of $(m - n - 1)$-sphere over M and is a Riemannian manifold of dimension $m - 1$ with the induced metric. We denote by dV_B the volume element of the bundle B. Let S_0^{m-1} be the unit hypersphere of E^m centered at the origin, $d\sigma$ the volume element of S_0^{m-1} with its induced metric from the euclidean metric $\tilde{g}$ of E^m, and

$$c_{m-1} = \int_{S_0^{m-1}} d\sigma \tag{2.1}$$

the volume of S_0^{m-1}. If we denote by ν the canonical map $B \to S_0^{m-1}$ which assigns each unit normal vector in B the unit vector in E^m through the origin of E^m and parallel to the normal vector, then by a simple direct computation, we have

$$\nu^* d\sigma = G(\xi) dV_B, \tag{2.2}$$

where $G(\xi) = \det A_\xi = M_n(\xi)$ for any unit normal vector ξ of M in E^m.

The *total absolute curvature* $\tau(x)$ of the immersion x, in the sense of Chern and Lashof (1957, 1958), is then defined by

$$\tau(x) = \frac{1}{c_{m-1}} \int_B |\nu^* d\sigma| = \frac{1}{c_{m-1}} \int_B |G(\xi)| dV_B. \tag{2.3}$$

Theorem 2.1 (Chern and Lashof, 1958). *Let $x: M \to E^m$ be an immersion of an n-dimensional closed manifold M in a euclidean m-space E^m. Then the total absolute curvature of x satisfies the following inequality:*

$$\tau(x) \geqq b(M). \tag{2.4}$$

Proof. For any unit vector $a \in S_0^{m-1}$ we define the height function h_a in the direction a by

$$h_a(P) = \tilde{g}(a, x(P)), \quad P \in M.$$

If ξ is a unit normal vector at P, that is, $(P, \xi) \in B$, then we have

$$dh_\xi(P) = \tilde{g}(\xi, dx(P)) = 0.$$

Hence, P is a critical point of the function h_ξ. Conversely, if P is a critical point of the height function h_a, $a \in S_0^{m-1}$, then we have

$$dh_a(P) = \tilde{g}(a, dx(P)) = 0.$$

This implies that a is a unit normal vector of M at P, that is, $(P, a) \in B$. Consequently, we see that the number of all critical points of h_a which is denoted by

$\beta(h_a)$ is equal to the number of points in M with a as its normal vector. Hence we have

$$\int_B |v^* d\sigma| = \int_{a \in S_0^{m-1}} \beta(h_a) d\sigma. \tag{2.5}$$

Since for each $a \in S_0^{m-1}$, h_a has a degenerate critical point if and only if a is a critical value of the map $v: B \to S_0^{m-1}$. By Sard's theorem the image of the set of critical points of v has measure zero in S_0^{m-1}. Hence for almost all $a \in S_0^{m-1}$, h_a is a nondegenerate function. Thus $\beta(h_a)$ is well-defined and is finite for almost all $a \in S_0^{m-1}$. Hence, by applying Theorem 1.1 to (2.5), we obtain the theorem. $\quad\square$

Theorem 2.2 (Chern and Lashof, 1957). *Let $x: M \to E^m$ be an immersion of an n-dimensional closed manifold M into a euclidean m-space E^m. If the total absolute curvature $\tau(x)$ of x satisfies the inequality*

$$\tau(x) < 3, \tag{2.6}$$

then M is homeomorphic to an n-sphere.

Proof. Suppose that the inequality (2.6) holds. Then there exists a set of positive measure on S_0^{m-1} such that, if a is a unit vector in this set, the height function h_a has just two critical points. For, if not, every point of S_0^{m-1}, except for a set of measure zero, would be covered at least three times by v and hence by (2.5) we would have $\tau(x) \geqq 3$. Since, by Sard's theorem, the image of the set of critical points under v is of measure zero, there is a unit vector a such that h_a has exactly two nondegenerate critical points. Thus, by applying Theorem 1.2 of Reeb, we see that M is homeomorphic to an n-sphere. This proves the theorem. $\quad\square$

For a hypersurface M in a euclidean $(n+1)$-space E^{n+1}, if for each point $P \in M$, the hyperplane (or great hypersphere) H_P of E^{n+1} tangent to M at P does not separate M into two parts, then M is called a *convex hypersurface* of E^{n+1}.

For an immersion of M with total absolute curvature 2 we have the following:

Theorem 2.3 (Chern and Lashof, 1957). *Let $x: M \to E^m$ be an immersion of an n-dimensional closed manifold M into a euclidean m-space E^m. Then $\tau(x) = 2$ if and only if the immersion x is an imbedding and $x(M)$ is a convex hypersurface in an $(n+1)$-dimensional linear subspace of E^m.*

For the proof of this theorem see Chern and Lashof (1957).

Remark 2.1. Theorems 2.1 and 2.3 generalized known results of Fenchel (1929) (case $m = 3$) and Borsuk (1947) (case $m > 3$) for closed curves in a euclidean m-space E^m, namely, if C is a closed curve in a euclidean m-space E^m with the position vector field $X = X(s)$, s being the arc length of C, then we have

$$\int_C |\kappa| ds \geq 2\pi, \tag{2.7}$$

where $|\kappa| = |d^2 x/ds^2|$. The equality sign of (2.7) holds when and only when C is a convex plane curve in E^m.

Remark 2.2. Let M be an n-dimensional orientable closed submanifold imbedded in a euclidean m-space E^m. Then we have

$$\int_B v^* d\sigma = \int_B G(\xi) dV_B = c_{m-1} \chi(M). \tag{2.8}$$

This result was proved independently by Allendoerfer (1940) and Fenchel (1940).

Remark 2.3. If n is even, then formula (2.8) can be written as follows (by a nontrivial proof)

$$\frac{c_n}{2} \chi(M) = \int_M \frac{1}{\mathfrak{g}} \{ \delta^{i_1 \cdots i_n} \delta^{j_1 \cdots j_n} K_{i_1 i_2 j_1 j_2} K_{i_3 i_4 j_3 j_4} \cdots K_{i_{n-1} i_n j_{n-1} j_n} \} dV, \tag{2.9}$$

where $\delta^{i_1 \cdots i_n}$ is zero if $i_1, \ldots, i_n$ do not form a permutation of $1, \ldots n$, and is equal to 1 or -1 according as the permutation is even or odd. Equation (2.9) is the well-known *formula of Gauss-Bonnet*. (See for instance, Chern, 1944a.)

3 Total Mean Curvature

Let $x \colon M \to E^m$ be an immersion of an n-dimensional closed manifold M in a euclidean m-space E^m. For a unit normal vector ξ of M at P let $M_i(\xi)$ denote the i-th mean curvature with respect to ξ. The i-th *total absolute curvature* $TA_i(x)$ is then defined by

$$TA_1(x) = \frac{1}{c_{m-1}} \int_B |M_i(\xi)|^{n/i} dV_B, \tag{3.1}$$

where dV_B is the volume element of the bundle B of all unit normal vectors of M in E^m. From (2.3) and (3.1) we see that the total absolute curvature $\tau(x)$ in the sense of Chern and Lashof is just the n-th total absolute curvature $TA_n(x)$.

Theorem 3.1. *Let $x \colon M \to E^m$ be an immersion of an n-dimensional closed manifold in a euclidean m-space E^m. Then we have*

$$TA_i(x) \geq 2, \tag{3.2}$$

for $i = 1, 2, \ldots, n - 1$.

Proof. For a fixed unit vector $a \in S_0^{m-1}$, the height function h_a has at least one maximum and one minimum, say Q and Q', respectively, since at (Q, a) and

(Q', a), the second fundamental tensors with respect to (Q, a) and (Q', a) are semidefinite. Hence, if we let U^* denote the set of all points $(P, \xi) \in B$ such that the second fundamental tensor with respect to (P, ξ) is semidefinite, then we know that the unit sphere S_0^{m-1} is covered by U^* under the mapping $v: B \to S_0^{m-1}$ at least twice. Since by (2.2) the integral

$$\int_{U^*} |G(\xi)| dV_B$$

is the volume of the image of U^* under v. Hence we have

$$\int_{U^*} |G(\xi)| dV_B \geq 2c_{m-1}. \tag{3.3}$$

On the other hand, by Problem II.2, we have

$$|M_i(\xi)|^n \geq |M_n(\xi)|^i = |G(\xi)|^i$$

for any unit normal vector ξ at P with $(P, \xi) \in U^*$. Hence, by (3.3), we obtain

$$TA_i(x) = (c_{m-1})^{-1} \int_B |M_i(\xi)|^{n/i} dV_B$$

$$\geq (c_{m-1})^{-1} \int_{U^*} |M_i(\xi)|^{n/i} dV_B$$

$$\geq (c_{m-1})^{-1} \int_{U^*} |G(\xi)| dV_B \geq 2. \tag{3.4}$$

This completes the proof of the theorem.   □

Theorem 3.2. *Let* $x: M \to E^m$ *be an immersion of an n-dimensional closed manifold M in a euclidean m-space E^m. Then the mean curvature vector H of M satisfies the inequality*

$$\int_M |H|^n dV \geq c_n, \tag{3.5}$$

where dV is the volume element of M.

Proof. Let $\xi_1, \ldots, \xi_{m-n}$ be $m - n$ mutually orthogonal unit normal vectors of M at P such that $H = |H|\xi_1$. Then we have

$$\text{trace } A_x = 0, \quad x = 2, \ldots, m - n, \tag{3.6}$$

$$(\text{trace } A_1)^2 = n^2 |H|^2. \tag{3.7}$$

Let ξ be any unit normal vector of M at P, we may write

$$\xi = \sum_x \cos \theta_x \xi_x,$$

where θ_x denotes the angle between ξ_x and ξ. Then we have

$$A_\xi = \sum_x \cos\theta_x A_x,$$

from which we find

$$|M_1(\xi)| = |\cos\theta_1||H|.$$

Hence we obtain

$$\int_B |M_1(\xi)|^n dV_B = \int_B |\cos\theta_1|^n |H|^n dV_B$$
$$= (2c_{m-1}/c_n) \int_M |H|^n dV, \qquad (3.8)$$

where we have used the following identity in spherical integration:

$$\int_{S_0^{m-1}} |\cos\theta|^n d\sigma = 2c_{n+m-1}/c_n, \qquad (3.9)$$

θ being the angle between the x^1-axis and the vector $a \in S_0^{m-1}$.

If the equality sign of (3.2) or (3.5) holds, then we may prove that the immersion x is an imbedding and M is a small n-sphere in E^m. For the proof of this fact see Chen (1971g). $\qquad \square$

Remark 3.1. If the dimension n of M is 2 and the codimension $m - n$ is 1, then Theorem 3.2 was first proved by T.J. Willmore (1968a).

Corollary 3.1. *Let M be an n-dimensional closed minimal submanifold of a unit hypersphere of E^m, then the volume of M, $v(M)$, satisfies the inequality*

$$v(M) \geqq c_n. \qquad (3.10)$$

The equality sign of (3.10) holds if and only if M is imbedded as a great n-sphere in the unit hypersphere of E^m.

If M is a minimal submanifold of a unit hypersphere of E^m, then the mean curvature vector H has constant length 1; hence, this corollary follows immediately from the above discussion.

Remark 3.2. If M is a closed surface in an n-dimensional euclidean space E^n, then the integral $\int_M |H|^2 dV$ is a conformal invariant (Chen, 1973b). For $n = 3$, see White (1973).

4 Submanifolds with Nonnegative Scalar Curvature

Let M be an n-dimensional closed submanifold of a euclidean m-space E^m. From §3 we know that the mean curvature $|H|$ of M satisfies the inequality

$$\int_M |H|^n dV \geqq c_n. \tag{4.1}$$

The equality sign of (4.1) holds when and only when M is imbedded as an n-sphere of E^m. It is interesting to know whether the inequality (4.1) can be improved for some special submanifolds of E^m.

The main purpose of this section is to study the integral of $|H|^n$ over a submanifold M and prove the following:

Theorem 4.1. *Let M be an n-dimensional closed submanifold of a euclidean m-space E^m with nonnegative scalar curvature. Then we have*

$$\int_M |H|^n dV > \gamma b(M), \tag{4.2}$$

where

$$\gamma = (4n^n)^{-1/2}c_n, \quad \textit{if n is even}$$
$$\gamma = (2n^n c_{m-n-1}c_{m+n-1})^{-1/2}(c_{2n})^{1/2}c_{m-1}, \quad \textit{if n is odd.} \tag{4.3}$$

We first prove the following lemmas.

Lemma 4.1. *Let $a_1, \ldots, a_p$ be p nonnegative constants and S_0^{p-1} be the unit hypersphere of a euclidean p-space E^p centered at the origin. Let f be the function on S_0^{p-1} defined by*

$$f(x) = \sum_{i=1}^{p} a_i(x_i)^2, \tag{4.4}$$

where $x = (x_1, \ldots, x_p)$, $x_1, \ldots, x_p$ the natural euclidean coordinates of E^p. If $2d$ is a positive even integer, then we have

$$\left(\sum_{i=1}^{p} a_i\right)^d \geqq (c_{2d})(2c_{p+2d-1})^{-1} \int_{S_0^{p-1}} \left(\sum_{i=1}^{p} a_i(x_i)^2\right)^d d\sigma, \tag{4.5}$$

where $d\sigma$ is the volume element of S_0^{p-1}. If the equality sign of (4.5) holds, then we have either $p = 1$ or $d = 1$ or at least $p - 1$ of $a_1, \ldots, a_p$ are zero. The converse of this is also true.

Proof. For nonnegative even integers $e_1, \ldots, e_p$, we have

$$\int_{S_0^{p-1}} (x_1)^{e_1} \cdots (x_p)^{e_p} d\sigma = \frac{2\Gamma((1+e_1)/2) \cdots \Gamma((1+e_p)/2)}{\Gamma((p+e_1+\cdots+e_p)/2)} \tag{4.6}$$

and

$$\Gamma((1+e_1)/2) \cdots \Gamma((1+e_p)/2) \leqq \Gamma((1+e_1+\cdots+e_p)/2)\Gamma(1/2)^{p-1}. \tag{4.7}$$

The equality sign of (4.7) holds when and only when at least $p-1$ of $e_1, \ldots, e_p$ are zero. From (4.6) and (4.7) we find

$$\int_{S_0^{p-1}} \left(\sum_{i=1}^p a_i(x_i)^2 \right)^d d\sigma \leqq \left(\sum_{i=1}^p a_i \right)^d \int_{S_0^{p-1}} (x_i)^{2d} d\sigma$$

$$= 2c_{p+2d-1}(c_{2d})^{-1} \left(\sum_{i=1}^p a_i \right)^d, \tag{4.8}$$

from which we obtain (4.5). If the equality sign of (4.5) holds, then the inequality in (4.8) is actually an equality. Hence we have either $d = 1$ or $p = 1$ or at least $p - 1$ of $a_1, \ldots, a_p$ are zero. The converse of this is trivial. $\qquad\square$

Lemma 4.2. *Let $a_1, \ldots, a_p$ be p nonnegative constants. If $2d$ is a positive odd integer and at least one of $a_1, \ldots, a_p$ is nonzero, then we have*

$$\left(\sum_{i=1}^p a_i \right)^d \geqq (c_{4d})^{1/2}(2c_{p-1}c_{4d+p-1})^{-1/2} \int_{S_0^{p-1}} \left(\sum_{i=1}^p a_i(x_i)^2 \right)^d d\sigma. \tag{4.9}$$

The equality sign of (4.9) holds when and only when $p = 1$.

This lemma follows immediately from Lemma 4.1 and Schwarz' inequality

$$\left| \int fg d\sigma \right| \leqq \left(\int f^2 d\sigma \right)^{\frac{1}{2}} \left(\int g^2 d\sigma \right)^{\frac{1}{2}},$$

where the equality sign holds when and only when $f = cg$ with $c = $ constant.

Lemma 4.3. *For each unit normal vector ξ of M at a point $P \in M$, we have*

$$(N(A_\xi))^n \geqq n^n(G(\xi))^2. \tag{4.10}$$

The equality sign of (4.10) holds when and only when $(A_\xi)^2 = cI_n$ for some constant c, where I_n is the $n \times n$ identity matrix.

Proof. Let $h_1, \ldots, h_n$ be the principal curvatures of M with respect to the normal direction ξ. Then we have

$$N(A_\xi) = \sum_{i=1}^{n}(h_i)^2 \geq n(h_1 \cdots h_n)^{2/n}$$
$$= n(G(\xi))^{2/n}.$$

Since $N(A_\xi)$ and $G(\xi)$ are both independent of the choice of the basis of the tangent space at P, we get (4.10).

If the equality sign of (4.10) holds, then we have

$$(h_1)^2 = \cdots = (h_n)^2. \tag{4.11}$$

This shows that $(A_\xi)^2 = cI_n$ for some constant c. Conversely, if we have $(A_\xi)^2 = cI_n$, then the principal curvatures $h_1, \ldots, h_n$ with respect to ξ satisfy (4.11). From this we obtain the equality sign of (4.10). This proves the lemma.   □

Lemma 4.4. *Let M be an n-dimensional submanifold of a euclidean m-space E^m and let S_p denote the $(m - n - 1)$-sphere of all unit normal vectors of M at a point $P \in M$. Then the length of second fundamental for $\langle h \rangle$ satisfies the following inequality*

$$\langle h \rangle^n \geq \begin{cases} c_n(2c_{m-1})^{-1} \int_{S_P} (N(A_\xi))^{n/2} dS_P, & \text{if } n \text{ is even,} \\ (c_{2n})^{\frac{1}{2}}(2c_{m-n-1}c_{m+n-1})^{-\frac{1}{2}} \int_{S_P}(N(A_\xi))^{\frac{n}{2}}dS_P, & \text{if } n \text{ is odd,} \end{cases} \tag{4.12}$$

where dS_P is the volume element of S_P. If the equality sign of (4.12) holds and n is even, then we have either $n = 2$ or there exist $m - n$ orthonormal vectors ξ_x such that $N(A_2) = \cdots = N(A_{m-n}) = 0$. If the equality sign of (4.12) holds and n is odd, then $m = n + 1$.

Proof. Let $\xi_1, \ldots, \xi_{m-n}$ be $m - n$ mutually orthogonal unit normal vector fields of M. Then we may write $\xi = \sum \cos \theta_x \xi_x$. Hence we have

$$N(A_\xi) = \sum_{x,y}(\text{trace } A_x A_y) \cos \theta_x \cos \theta_y. \tag{4.13}$$

The right-hand side of (4.13) is a quadratic form in the $\cos \theta_x$. Hence we can choose a local frame field $\bar{\xi}_1, \ldots, \bar{\xi}_{m-n}$ such that, with respect to this frame field, we have

$$N(A_\xi) = \sum \Phi_x \cos^2 \theta_x,$$
$$\Phi_1 \geq \Phi_2 \geq \cdots \geq \Phi_{m-n} \geq 0, \tag{4.14}$$
$$\Phi_x = \text{trace } A_x^2 = N(A_x). \tag{4.15}$$

By applying Lemmas 4.1 and 4.2 we obtain (4.12). If n is even and the equality sign of (4.12) holds, then, by Lemma 4.1, we have either $n = 2$ or

$\Phi_2 = \cdots = \Phi_{m-n} = 0$. If n is odd and the equality sign of (4.12) holds, then, by Lemma 4.2, we have $m = n + 1$. $\qquad\square$

Now, *we return to the proof of the theorem.* Let M be an n-dimensional closed submanifold of E^m with nonnegative scalar curvature r. Then, by Eq. (II.4.3), we have

$$n^2|H|^2 \geqq \langle h \rangle^2. \tag{4.16}$$

The equality sign of (4.16) holds when and only when the scalar curvature $r = 0$. From (2.3), (4.10), (4.12), and (4.16) we find

$$\int_M |H|^n dV \geqq \begin{cases} c_n(2^{-1})(n)^{n/2}\tau(x), & \text{if } n \text{ is even,} \\ c_{m-1}(c_{2n})^{\frac{1}{2}}(2n^n c_{m-n-1}c_{m+n-1})^{-\frac{1}{2}}\tau(x), & \text{if } n \text{ is odd,} \end{cases} \tag{4.17}$$

where x is the immersion of M in E^m. Thus, by applying Theorem 2.1 of Chern and Lashof, we find

$$\int_M |H|^n dV \geqq \gamma b(M). \tag{4.18}$$

Now, suppose that the equality sign of (4.18) holds. Then the inequalities (4.12) and (4.16) are actually equalities. Hence, the scalar curvature $r = 0$. If $n = 2$, then M is flat. Hence M is either a flat torus or a flat Klein bottle. Thus we have $b(M) = 4$. Hence, by the equality of (4.18), we find $\int_M |H|^2 dV = c_2$. This is impossible by virtue of Theorem 3.2. Thus, we have $n \neq 2$. Hence, by Lemma 7.4, we see that, for each point $P \in M$, there exist $m - n$ orthonormal normal vectors $\xi_1, \ldots, \xi_{m-n}$ of M at P such that

$$A_2 = \cdots = A_{m-n} = 0. \tag{4.19}$$

From (4.19), we find that $M_2(\xi_2) = \cdots = M_2(\xi_{m-n}) = 0$ and $0 = r = (\text{trace } A_1)^2 - \text{trace}(A_1)^2$. On the other hand, from the definition of the second mean curvature $M_2(\xi_1)$ of ξ_1, we have

$$\binom{n}{2} M_2(\xi_1) = (\text{trace } A_1)^2 - \text{trace}(A_1)^2.$$

Hence we see that the second mean curvature $M_2(\xi_1)$ vanishes everywhere. This is impossible by virtue of Theorem 3.1. Hence, we obtain the theorem. $\qquad\square$

Remark 4.1. From Theorem 4.1 we see that if M is a closed submanifold of E^m with nonnegative scalar curvature, then the integral $\int_M |H|^n dV$ depends on the topological structure of M. In particular, if $b(M)$ is large, the $\int_M |H|^n dV$ is large.

Theorem 4.2. *Let M be an n-dimensional closed submanifold of a euclidean m-space E^m with nonnegative scalar curvature. Then we have*

$$\int_M \langle h \rangle^n dV \geqq \frac{1}{2} n^{\frac{n}{2}} c_n b(M) \tag{4.20}$$

for even n and

$$\int_M \langle h \rangle^n dV \geqq \frac{(n^n c_{2n})^{1/2} c_{m-1}}{(2c_{m-n-1} c_{m+n-1})^{1/2}} b(M) \tag{4.21}$$

for odd n. The equality sign of (4.20) holds when and only when M is imbedded as a small n-sphere of E^m, and the equality sign of (4.21) holds when and only when $m = n + 1$ and M is imbedded as a small hypersphere of E^{n+1}.

Proof. Inequalities (4.20) and (4.21) follow immediately from Lemmas 4.3 and 4.4 and Theorem 2.1. If the equality signs of (4.20) and (4.21) hold, then inequalities (4.10) and (4.12) are actually equalities. Hence, from Lemma 4.3, we have $(A_\xi)^2 = f I_n$ for some function f and every unit normal vector ξ. Moreover, we have either $n = 2$ or (4.19) for suitable ξ_x at $P \in M$ for each P.

Case (i). If $n = 2$, then, for each unit normal vector ξ of M at a point $P \in M$, the second fundamental tensor A_ξ is one of the following forms:

$$\begin{pmatrix} a & 0 \\ 0 & a \end{pmatrix} \quad \text{or} \quad \begin{pmatrix} b & b \\ c & -b \end{pmatrix}. \tag{4.22}$$

Now, suppose that, at some point $P \in M$, there exist unit normal vectors ξ and ξ' of M at P such that

$$A_\xi = \begin{pmatrix} a & 0 \\ 0 & a \end{pmatrix} \quad \text{and} \quad A_{\xi'} = \begin{pmatrix} b & c \\ c & -b \end{pmatrix}$$

with $a \neq 0$ and at least one of b, c being nonzero. Let $\bar{\xi} = \cos \theta \xi + \sin \theta \xi'$. Then we have

$$A_{\bar{\xi}} = \begin{pmatrix} a \cos \theta + b \sin \theta & c \sin \theta \\ c \sin \theta & a \cos \theta - b \sin \theta \end{pmatrix}$$

Since $(A_{\bar{\xi}})^2 = c I_2$ for some c, we find

$$ab \sin 2\theta = 0, \quad ac \sin 2\theta = 0. \tag{4.23}$$

Since this is true for all θ, (4.23) implies that $b = c = 0$. This is a contradiction. Thus, we see that, for each point $P \in M$, the second fundamental tensor A_ξ is either in the form

$$\begin{pmatrix} \lambda(\xi) & 0 \\ 0 & \lambda(\xi) \end{pmatrix}$$

for all unit normal vectors at P, or in the form

$$\begin{pmatrix} \lambda(\xi) & \mu(\xi) \\ \mu(\xi) & -\lambda(\xi) \end{pmatrix}$$

for all unit normal vectors at P. From Theorem 3.2 we see that the subset $U = \{P \in M; A_\xi = \lambda(\xi)I_2 \text{ for some } \lambda \neq 0 \text{ at } P\}$ is nonempty. Since U is totally umbilical in E^m, Proposition II.3.2 implies that each component of U is contained in a small 2-sphere in E^m. Thus the mean curvature $|H|$ of M is a constant on each component of U. By the continuity of the mean curvature on M, we see that $U = M$. This shows that M is a small 2-sphere of E^m.

Case (ii). If, for each point $P \in M$, there exist $m - n$ orthonormal normal vectors ξ_x at P such that

$$A_2 = \cdots = A_{m-n} = 0, \tag{4.24}$$

then, from the fact that $A_\xi^2 = f I_n$ and the inequality (3.3), we see that the subset $U = \{P \in M; A_1 = c I_n \text{ for some } c \neq 0 \text{ at } P\}$ is nonempty. Since U is totally umbilical in E^m, Proposition II.3.2 implies that each component of U is contained in a small n-sphere of E^m. Thus c is a constant on each component of U. Since $c = |H|$ and the mean curvature H is continuous on M, we see that $U = M$. Thus we have proved that the submanifold M is contained in a small n-sphere of E^m and M is homeomorphic to an n-sphere. Thus we have $b(M) = 2$. Substituting this into (4.20) and (4.21) we find that if the equality of (4.20) holds, then M is imbedded as a small n-sphere of E^m, and if the equality of (4.21) holds, then $m = n + 1$ and M is imbedded as a small hypersphere of E^m. This completes the proof of the theorem. $\qquad\square$

In the remaining part of this section we assume that M is an n-dimensional minimal submanifold of a unit hypersphere S^{m-1} of E^m and $\langle \bar{h} \rangle$ is the length of the second fundamental form of M in S^{m-1}. Then, by a direct computation, we have

$$\langle h \rangle^2 = n + \langle \bar{h} \rangle^2, \quad r = n(n-1) - \langle \bar{h} \rangle^2. \tag{4.25}$$

Theorem 4.3. *Let M be an n-dimensional closed minimal submanifold of a unit hypersphere S^{m-1} of E^m. If a is a nonnegative real number $\leq n(n-1)$ and if the volume $v(M)$ of M satisfies*

$$v(M) \leq \frac{n^{n/2}c_n}{2(a+n)^{n/2}}b(M) \tag{4.26}$$

for even n or

$$v(M) \leqq \frac{(n^n c_{2n})^{1/2} c_{m-1}}{(2c_{m-n-1}c_{m+n-1}(a+n)^n)^{1/2}} b(M) \qquad (4.27)$$

for odd n, then we have $\langle \overline{h} \rangle^2 > a$ *at some point of M unless* $a = 0$ *and M is totally geodesic.*

Proof. If the inequalities of (4.26) and (4.27) are strict inequalities, then this theorem follows immediately from Theorem 4.2. Now, suppose that the inequalities (4.26) and (4.27) are actually equalities and we have $\langle \overline{h} \rangle^2 \leqq a$ everywhere. Then the equality signs of (4.20) and (4.21) hold. Therefore, by Theorem 4.2, we see that M is totally geodesic in S^{m-1}. Hence, we have $b(M) = 2$. Substituting this into (4.26) and (4.27), we obtain $a = 0$. This proves the theorem. □

If we choose $a = n(n-1)$, then we have the following:

Corollary 4.1. *Let M be an n-dimensional closed minimal submanifold of a unit hypersphere* S^{m-1} *of* E^m. *If we have*

$$v(M) \leqq \frac{c_n}{2n^{n/2}} b(M) \qquad (4.28)$$

for even n or

$$v(M) \leqq \frac{(c_{2n})^{1/2} c_{m-1}}{(2n^n c_{m-n-1}c_{m+n-1})^{1/2}} b(M) \qquad (4.29)$$

for odd n, then $\langle \overline{h} \rangle^2 > n(n-1)$ *at some points of M.*

This corollary follows immediately from Theorem 4.3.

This corollary gives a partial answer for a problem proposed by Chern, do Carmo, and Kobayashi (1970), namely,

Problem 4.1. Let M be an n-dimensional closed minimal submanifold of a unit $(m-1)$-sphere with second fundamental form $\overline{h}$ of constant length $\langle \overline{h} \rangle$. What is the next possible value for $\langle \overline{h} \rangle^2$ greater than $n/[2 - 1/(m-n-1)]$?

5   Surfaces in a Euclidean 4-Space

The main purpose of this section is to improve inequality (4.1) for surfaces in a euclidean 4-space.

Let $x: M \to E^m$ be an immersion of a surface M in a euclidean m-space E^m. We choose $m-2$ mutually orthogonal unit normal vector fields $\xi_1, \ldots, \xi_{m-2}$

of M in E^m. Let ξ be a unit normal vector field given by

$$\xi = \sum \cos \theta_x \xi_x. \tag{5.1}$$

Then we have

$$A_\xi = \sum \cos \theta_x A_x, \tag{5.2}$$

from which, we find that

$$G(\xi) = \det \left(\sum_x \cos \theta_x A_x \right). \tag{5.3}$$

The right-hand side of (5.3) is a quadratic form of $\cos \theta_1, \ldots, \cos \theta_{m-2}$. Thus we may choose a suitable local orthonormal normal frames $\bar{\xi}_1, \ldots, \bar{\xi}_{m-2}$ such that, with respect to the $\bar{\xi}_x$, we have

$$G(\xi) = \sum \lambda_x \cos^2 \theta_x, \quad \lambda_1 \geq \lambda_2 \geq \cdots \geq \lambda_{m-2}. \tag{5.4}$$

From (5.4) we find

$$\lambda_x = G(\bar{\xi}_x), \tag{5.5}$$
$$r = 2(\lambda_1 + \cdots + \lambda_{m-2}), \tag{5.6}$$

where r is the scalar curvature of M. We call such a frame $\bar{\xi}_1, \ldots, \bar{\xi}_{m-2}$ an Ōtsuki frame (Ōtsuki, 1966; Chen, 1970a). We call λ_x the x-th curvature of the surface M in E^m. By means of the method of definition, the λ_x are defined continuously on the whole surface M and they are differentiable on the open subset on which $\lambda_1 > \lambda_2 > \cdots > \lambda_{m-2}$.

Lemma 5.1. *Let M be a surface in E^m. Then the $(m-2)$-th curvature λ_{m-2} is nonpositive everywhere on M for $m > 3$.*

Proof. If the mean curvature vector H vanishes at a point $P \in M$, then we have $G(\xi) = -\frac{1}{2} N(A_\xi)$ for any unit normal vector ξ of M at P; in particular, we have $\lambda_{m-2} \leq 0$ at P. If the mean curvature vector H is nonzero at a point P and if ξ is any unit normal vector of M at P which is perpendicular to H, then we have $G(\xi) = -\frac{1}{2} N(A_\xi)$; in particular, we have $\lambda_{m-2} \leq 0$ at P. $\qquad \square$

Lemma 5.2. *Let M be a closed surface in E^m with $\lambda_{m-2} = 0$. Then M is homeomorphic to a 2-sphere.*

Proof. Since $\lambda_{m-2} = 0$, we have $G(\xi) \geq 0$ everywhere. Hence, by (5.4) and (5.6), we find

$$
\begin{aligned}
c_{m-1}\tau(x) &= \int_B |v^*d\sigma| = \int_B |G(\xi)|dV_B \\
&= \int_B G(\xi)dV_B \\
&= \int_B (\lambda_1 \cos^2 \theta_1 + \cdots + \lambda_{m-2} \cos^2 \theta_{m-2})dV_B \\
&= \frac{c_{m-1}}{2\pi} \int_M (\lambda_1 + \cdots + \lambda_{m-2})dV \\
&= \frac{c_{m-1}}{4\pi} \int_M rdV,
\end{aligned} \tag{5.7}
$$

where x is the immersion of M in E^m.

Since the scalar curvature r is equal to twice the Gaussian curvature and the integral of the Gaussian curvature gives the Gauss-Bonnet formula which holds for orientable closed surfaces as well as nonorientable ones,

$$
\int_M \frac{r}{2}dV = 2\pi\chi(M), \tag{5.8}
$$

and from (5.7), we find

$$
\tau(x) = \chi(M). \tag{5.9}
$$

Equation (5.9) implies that the Euler characteristic of M is equal to two. Hence, we see that the surface M is homeomorphic to a 2-sphere. $\qquad\square$

Lemma 5.3. *Let M be a closed surface in E^m. Then the subset $U = \{P \in M;$ $\lambda_1 > 0$ at $P\}$ is a nonempty open subset of M.*

Proof. Let x be the immersion of M in E^m. If we have $\lambda_1 \leq 0$ everywhere on M, then we have

$$
c_{m-1}\tau(x) = -\int_B G(\xi)dV_B = -c_{m-1}\chi(M). \tag{5.10}
$$

Combining (5.10) and (2.4) we find $-\chi(M) \geq b(M)$. This is impossible. Thus we obtain the lemma. $\qquad\square$

Lemma 5.4. *Let M be a surface in E^m. Then we have*

$$
|H|^2 \geq \lambda_1 \tag{5.11}
$$

on the subset $V = \{P \in M : \lambda_1(P) \geq 0\}$. The equality sign of (5.11) holds everywhere on V when and only when V is pseudoumbilical in E^m.

Proof. Let $\xi_1, \ldots, \xi_{m-2}$ be an Ōtsuki frame. Then we have $\lambda_x = G(\xi_x)$, $x = 1, 2, \ldots, m-2$. Thus we find

$$
\begin{aligned}
4|H|^2 &= \sum (\text{trace } A_x)^2 = \sum N(A_x) + r \\
&\geqq 2(\lambda_1 - \lambda_2 - \cdots - \lambda_{m-2}) + 2(\lambda_1 + \lambda_2 + \cdots + \lambda_{m-2}) = 4\lambda_1. \quad (5.12)
\end{aligned}
$$

The inequality in (5.12) is an equality when and only when

$$
A_1 = |H|I_2, \quad \text{trace } A_x = 0, x = 2, \ldots, m-2. \quad (5.13)
$$

From (5.12) we obtain (5.11). If the equality sign of (5.11) holds, then the equality sign of (5.12) holds. Thus we have (5.13). From this we see that the subset V is pseudoumbilical in E^m. Conversely, if the subset V is pseudoumbilical in E^m and $\xi_1, \ldots, \xi_{m-2}$ are $m-2$ mutually orthogonal unit normal vector fields of M in E^m such that $H = |H|\xi_1$, then we have (5.13). From (5.13) we can easily verify that $|H|^2 = \lambda_1$ on V. This proves the lemma. $\quad\square$

In the remaining part of this section we always assume that $x: M \to E^4$ is an immersion of a closed surface in a euclidean 4-space E^4. We also simply denote by λ the first curvature λ_1 of M and the secondary curvature λ_2 of M by μ.

Theorem 5.1. *Let M be a closed surface in a euclidean 4-space with nonnegative Gaussian curvature. If the mean curvature of M satisfies the inequality*

$$
\int_M |H|^2 dV \leqq (2 + \pi)\pi, \quad (5.14)
$$

then M is homeomorphic to a 2-sphere.

Proof. Let x be the immersion of M in E^4. By the assumption, the Gaussian curvature is nonnegative, we have $\lambda \geq 0$ everywhere on M. Hence, by Lemma 5.4, we have $|H|^2 \geqq \lambda$. Since the scalar curvature is also nonnegative, by Lemma 5.1, we have

$$
\begin{aligned}
|G(\xi)| &= |(\lambda + \mu)\cos^2\theta - \mu\cos 2\theta| \\
&\leqq \frac{r}{2}\cos^2\theta - \mu|\cos 2\theta|. \quad (5.15)
\end{aligned}
$$

Hence, by (5.15), we find

$$
\begin{aligned}
2\pi^2\tau(x) &\leqq \int_B \left\{ \frac{r}{2}\cos^2\theta - \mu|\cos 2\theta| \right\} dV_B \\
&= 4\int_M \lambda dV - 4\left(1 - \frac{\pi}{4}\right)\int_M \frac{r}{2} dV. \quad (5.16)
\end{aligned}
$$

Thus, by applying to (5.16), Theorem 2.1 of Chern and Lashof and the formula of Gauss and Bonnet, we find

$$\int_M |H|^2 dV \geq \int_M \lambda \, dV$$

$$\geq \frac{\pi^2}{2} \{b(M) - \chi(M)\} + 2\pi \chi(M). \qquad (5.17)$$

Since the Gaussian curvature is nonnegative, M is one of the following surfaces: a topological 2-sphere, a real projective plane, or a flat surface.

If M is flat, then by Theorem 5.2, we have $\int_M |H|^2 dV \geq 2\pi^2$. This is a contradiction. Thus, M is either a topological 2-sphere or a real projective plane. If M is a topological 2-sphere, there is nothing to prove. Hence, we may assume that M is a real projective plane. In this case, we have $b(M) = 3$ and $\chi(M) = 1$. Thus, by (5.17), we find

$$\int_M |H|^2 dV \geq (2+\pi)\pi. \qquad (5.18)$$

If the equality sign of (5.18) holds, then the inequality in (5.15) is actually an equality for all unit normal vector ξ of M in E^4. Thus we obtain $\mu = 0$ on M. This is impossible by virtue of Lemma 5.2. Thus we obtain the theorem. □

Theorem 5.2. *Let M be a closed surface in a euclidean 4-space E^4 with nonnegative Gaussian curvature. Then we have*

$$\int_M |H|^2 dV \geq 2\pi^2. \qquad (5.19)$$

If the mean curvature of M is constant, then the equality sign of (5.19) holds when and only when M is imbedded as a Clifford torus in E^4, that is, M is the product surface of two plane circles with the same radii.

Proof. Let x denote the immersion of M in E^4 and let $p: B \to M$ be the projection of the bundle B onto M given by $p((P, \xi)) = P$. Let $W = \{P \in M; \lambda(P) > 0\}$. Then, by the assumption that the Gaussian curvature is nonpositive, we have

$$|G(\xi)| = |\lambda \cos^2 \theta + \mu \sin^2 \theta|$$

$$= \left| \lambda \cos 2\theta + \frac{r}{2} \sin^2 \theta \right|$$

$$\leq \lambda |\cos 2\theta| - \frac{r}{2} \sin^2 \theta, \qquad (5.20)$$

on the subset $p^{-1}(W)$. Thus we have

$$\int_{p^{-1}(W)} |G(\xi)| dV_B \leq 4 \int_W \lambda \, dV - \frac{\pi}{2} \int_W r \, dV. \qquad (5.21)$$

On the other hand, on the subset $B - p^{-1}(W)$ we have

$$\int_{B-p^{-1}(W)} |G(\xi)| dV_B = -\int_{B-p^{-1}(W)} G(\xi) dV_B$$

$$= -\frac{\pi}{2} \int_{M-W} r \, dV. \tag{5.22}$$

Thus, by applying Lemma 5.4, we find

$$\int_M |H|^2 dV \geqq \int_W \lambda \, dV$$

$$\geqq \frac{1}{4} \int_{p^{-1}(W)} |G(\xi)| dV_B + \frac{\pi}{8} \int_W r \, dV$$

$$+ \frac{1}{4} \int_{B-p^{-1}(W)} |G(\xi)| dV_B + \frac{\pi}{8} \int_{M-W} r \, dV$$

$$= \frac{\pi^2}{2} \tau(x) + \frac{\pi^2}{2} \chi(M)$$

$$\geqq \frac{\pi^2}{2} \{b(M) + \chi(M)\} = 2\pi^2. \tag{5.23}$$

This proves the inequality (5.19). Now, suppose the mean curvature of M is constant. If the equality sign of (5.19) holds, then the inequalities in (5.23) are actually equalities. From this we see that λ is a positive constant on M and, by Lemma 5.4, we see that M is pseudoumbilical in E^4. From this we know that M is contained in a small hypersphere of E^4 as a minimal surface. Since the Gaussian curvature of M is nonpositive everywhere, by applying Problem III.5, we see that M is immersed as a Clifford torus in E^4. Since, for a Clifford torus in E^4, we have $\int_M |H|^2 dV = 2\pi^2$, thus we see that x is an imbedding. This completes the proof of the theorem. $\square$

Theorem 5.3. *Let $x: M \to E^4$ be an immersion of a closed surface in E^4 with nonpositive Gaussian curvature. Then we have*

$$\int_M |\lambda| dV \geqq 2\pi^2. \tag{5.24}$$

The equality sign of (5.24) holds when and only when M is flat and $\tau(x) = b(M)$.

Proof. Let $W = \{P \in M; \lambda(P) > 0\}$. Then, by (5.23), we have

$$\int_M |\lambda| dV \geqq \int_W \lambda \, dV \geqq \frac{\pi}{2} \{\tau(x) + \chi(M)\}$$

$$\geqq \frac{\pi^2}{2} \{b(M) + \chi(M)\} = 2\pi^2. \tag{5.25}$$

This proves (5.24). If the equality sign of (5.24) holds, then we have $\tau(x) = b(M)$, $\lambda \geq 0$ and

$$
\begin{aligned}
|G(\xi)| &= \left| \lambda \cos 2\theta + \frac{r}{2} \sin^2 \theta \right| \\
&= \lambda |\cos 2\theta| - \frac{r}{2} \sin^2 \theta
\end{aligned}
\tag{5.26}
$$

for all unit normal vector $\xi = \cos \theta \xi_1 + \sin \theta \xi_2$, where $\{\xi_1, \xi_2\}$ is an Ōtsuki frame. Thus we see that $r = 0$, that is, M is flat. Conversely, if M is flat and $\tau(x) = b(M)$, then we have (5.26) and $\tau(x) = b(M)$. From this we can easily verify that the inequalities in (5.25) are actually equalities. Hence we get the equality sign of (5.24). This proves the theorem.   □

Remark 5.1. If M is a flat torus in E^4, then Theorem 5.3 was obtained by Ōtsuki (1966) (see also Chen and Houh, 1972).

6   Tori in a Euclidean 3-Space

In the previous section we proved that if M is a closed surface in a euclidean 4-space with nonpositive Gaussian curvature, then the integral $\int_M |H|^2 dV$ is always $\geq 2\pi^2$. In this section, we shall consider surface in a euclidean 3-space and prove the following theorem of Shiohama, Takagi, and Willmore.

Theorem 6.1 (Shiohama and Takagi, 1970; Willmore, 1971a). *Let $x: M \to E^3$ be an imbedding of a torus in a euclidean 3-space E^3. If $x(M)$ is the surface generated by carrying a small circle around a closed curve so that the center moves along the curve and the plane of the circle is in the normal plane to the curve at each point, then we have*

$$
\int_M |H|^2 dV \geq 2\pi^2.
\tag{6.1}
$$

The equality sign of (6.1) holds when and only when $x(M)$ is congruent to the anchor-ring in E^3 with the euclidean coordinates $\{x^1, x^2, x^3\}$ given by

$$
\begin{aligned}
x^1 &= (\sqrt{2}a + a \cos u) \cos v, \\
x^2 &= (\sqrt{2}a + a \cos u) \sin v, \\
x^3 &= a \sin u,
\end{aligned}
\tag{6.2}
$$

where a is a positive constant.

Proof. Let C be the closed curve mentioned in the theorem and the position vector field of C in E^3 be given by $X = X(s)$, where s is the arc length of C. Let κ

and τ be the *curvature* and *torsion* of the curve C in E^3. Then the formulas of Frenet-Serret are given by

$$\frac{dX}{ds} = E,$$

$$\frac{dE}{ds} = \kappa\zeta,$$

$$\frac{d\zeta}{ds} = -\kappa E + \tau\eta,$$

$$\frac{d\eta}{ds} = -\tau\zeta, \tag{6.3}$$

where E is a unit tangent vector field of C, and ζ, η are two unit normal vector fields of C such that η is the vector product of E and ζ, that is, $\eta = E \times \zeta$.

Let Y denote the position vector field of the surface M in E^3. Then, by the hypothesis of the theorem, there exists a positive number c such that

$$Y(s, v) = X(s) + c\cos v\zeta + c\sin v\eta. \tag{6.4}$$

By a straightforward computation we find that the principal curvatures h_1, h_2 of the surface M in E^3 are given by

$$h_1 = \frac{1}{c}, \quad h_2 = \frac{\kappa\cos v}{\kappa c\cos v - 1},$$

from which we find that the mean curvature vector H of M satisfies

$$|H|^2 = \left\{\frac{1 - 2\kappa c\cos v}{2c(1 - \kappa c\cos v)}\right\}^2. \tag{6.5}$$

Thus (6.5) implies

$$\int_M |H|^2 dV = \int_0^l \int_0^{2\pi} \left\{\frac{1 - 2\kappa c\cos v}{2c(1 - \kappa c\cos v)}\right\}^2 ds \wedge dv$$

$$= \frac{\pi}{2c} \int_0^l (1 - \kappa^2 c^2)^{-1/2} ds, \tag{6.6}$$

where l is the length of the curve C. From (6.6) we find

$$\int_M |H|^2 dV = \frac{\pi}{2} \int_0^l \frac{|\kappa|}{|\kappa c|\sqrt{1 - \kappa^2 c^2}} ds$$

$$\geq \pi \int_0^l |\kappa| ds, \tag{6.7}$$

by virtue of the fact that, for any real variable x, the expression $x\sqrt{(1 - x^2)}$ takes its maximum value $\frac{1}{2}$ when $x = 1/\sqrt{2}$.

By applying Fenchel's theorem, inequality (2.7), to (6.7), we obtain (6.1). Now, assume that the equality sign of (6.1) holds. Then the inequality in (6.7) is actually an equality and the equality sign of (2.7) holds. Hence, we find that C is a convex plane closed curve in E^3 with the curvature $\kappa = (2c^2)^{-1/2}$. From this we can verify that C is a circle with radius $\sqrt{2}c$. This proves that M is imbedded as an anchor-ring of the type given by (6.2). The converse of this is trivial.   □

Remark 6.1. Shiohama and Takagi (1970) prove that if M is a closed oriented surface imbedded in a euclidean 3-space E^3 with one constant principal curvature and if M is not of genus zero, then M is of genus one and it is imbedded as a surface given by the assumption of Theorem 6.1, that is, M is a surface generated by carrying a small circle around a closed curve so that the center moves along the curve and the plane of the circle is in the normal plane to the curve at each point.

Remark 6.2. T.J. Willmore (1968a) conjectured that inequality (6.1) is true for all tori in E^3.

7 Stable Hypersurfaces

Let M be an n-dimensional oriented hypersurface in a euclidean $(n+1)$-space E^{n+1} with ξ as the unit outer normal vector field and H as the mean curvature vector. The function α defined by

$$H = \alpha\xi \qquad (7.1)$$

is a well-defined smooth function on M.

Let $\tilde{X}$ denote the position vector field of M in E^{n+1} with respect to the origin of E^{n+1}. Then the hypersurface M can be characterized by the vector-valued function

$$\tilde{X} = \tilde{X}(u^1, \ldots, u^n), \qquad (7.2)$$

where $u^1, \ldots, u^n$ define a local coordinate system in M. Since M is oriented, we may assume that $\partial/\partial u^1, \ldots, \partial/\partial u^n$ define the orientation of M. If we put $X_i = \partial\tilde{X}/\partial u^i$, then we have

$$g_{ji} = \langle X_j, X_i \rangle, \qquad (7.3)$$

$$\xi = \frac{[X_1, \ldots, X_n]}{\sqrt{\mathfrak{g}}}, \quad \mathfrak{g} = \det(g_{ji}), \qquad (7.4)$$

$$h_{ji} = -\langle \xi_j, X_i \rangle = h_{ij}, \quad \xi_j = \partial\xi/\partial u^j, \qquad (7.5)$$

where $\langle \, , \, \rangle$ is the scalar product in E^{n+1} and h_{ji} the second fundamental form of M in E^{n+1}.

Let $\varphi = \varphi(u^1, \ldots, u^n)$ be a differentiable function on M and t an infinitesimal. We put

$$\overline{X}(u^1, \ldots, u^n, t) = \tilde{X}(u^1, \ldots, u^n) + t\varphi\xi(u^1, \ldots, u^n). \qquad (7.6)$$

Then (7.6) defines a family of hypersurfaces in E^{n+1} nearby M. If $\varphi = 0$ on ∂M, then (7.6) is called a *normal variation* of M in E^{n+1}.

In this section we only consider the normal variations which leave ∂M *strongly fixed* in the sense that both φ and its gradient vanish identically on ∂M. If M is closed, then ∂M is empty and there is no restriction on the normal variations.

In the following, we simply denote by δ the operator $\partial/\partial t|_{t=0}$. Then, by a direct simple computation, we have

$$\delta\tilde{X} = \varphi\xi, \quad \partial X_i = \varphi_i\xi + \varphi\xi_i, \qquad (7.7)$$

$$\delta g_{ij} = -2\varphi h_{ij}, \quad \delta g^{ij} = 2\varphi h^{ij}, \qquad (7.8)$$

$$\delta\sqrt{\mathfrak{g}} = -n\alpha\varphi\sqrt{\mathfrak{g}}, \quad n\alpha = g^{ji}h_{ji}, \qquad (7.9)$$

where $h^{ij} = g^{it}g^{js}h_{ts}$.

From the equation (II.1.10) of Gauss and the equation (II.1.11) of Weingarten, we find

$$X_{ij} = \Gamma^h_{ij}X_h + h_{ij}\xi, \quad X_{ij} = \frac{\partial^2\tilde{X}}{\partial u^i\partial u^j}, \qquad (7.10)$$

$$\xi_i = -h^j_i X_j, \quad h^j_i = g^{jt}h_{ti}, \qquad (7.11)$$

$$\langle \xi_i, \xi \rangle = 0. \qquad (7.12)$$

Thus, by using (7.11) and (7.12), we find

$$\delta X_{ij} \equiv \varphi_{ij}\xi + \varphi\xi_{ij} \pmod{X_l}$$
$$\equiv \varphi_{ij}\xi - \varphi h^k_i X_{kj} \pmod{X_l}$$
$$\equiv \varphi_{ij}\xi - \varphi h^k_i h_{kj}\xi \pmod{X_l}. \qquad (7.13)$$

Hence, we obtain

$$\langle \xi, \delta X_{ij} \rangle = \varphi_{ij} - \varphi h^k_i h_{kj}. \qquad (7.14)$$

Thus, from (7.4), (7.7), (7.9), (7.10), (7.11), and the identity

$$\langle J, [A_1, \ldots, A_n] \rangle = (-1)^n \det(J, A_1, \ldots, A_n), \qquad (7.15)$$

for any $n+1$ vectors in E^{n+1}, we find

$$\langle X_{ij}, \delta\xi \rangle = -\varphi_k\Gamma^k_{ij}. \qquad (7.16)$$

Thus, by (7.14) and (7.16), we find

$$\delta h_{ij} = \varphi_{ij} - \varphi_k\Gamma^k_{ij} - \varphi h^k_i h_{kj}. \qquad (7.17)$$

From (7.17) we obtain

$$n(\delta\alpha) = \Delta\varphi + \varphi h^i_j h^j_i, \tag{7.18}$$

where $\Delta\varphi$ is the Laplacian of φ on M. Equation (7.18) implies that α and the scalar curvature r of M satisfy

$$n(\delta\alpha) = \Delta\varphi + n^2\varphi\alpha^2 - r\varphi. \tag{7.19}$$

Thus, from (7.9) and (7.19), we find

$$\delta \int_M \alpha^c dV = \int_M \left\{ \frac{c}{n}\alpha^{c-1}\Delta\varphi + n(c-1)\alpha^{c+1}\varphi - \frac{c}{n}\varphi\alpha^{c-1}r \right\} dV \tag{7.20}$$

for any $c \geq 0$.

If we integrate by parts to get rid of the derivatives of φ, we see that

$$\int_M \alpha^{c-1}\Delta\varphi dV = \int_M (\Delta\alpha^{c-1})\varphi dV, \tag{7.21}$$

where the boundary terms one would expect after integration by parts vanish because of our hypothesis of the vanishing of φ and the gradient of φ on the boundary ∂M.

Combining (7.20) and (7.21) we find

$$\delta \int_M \alpha^c dV = \int_M \varphi \left\{ \frac{c}{n}\Delta\alpha^{c-1} + n(c-1)\alpha^{c+1} - \frac{c}{n}\alpha^{c-1}r \right\} dV. \tag{7.22}$$

In the following, by an *S-hypersurface* in E^{n+1} we mean a closed oriented hypersurface M of E^{n+1} which is stable with respect to the integral $\int_M \alpha^n dV$, that is, for any normal variation of M in E^{n+1}, we have $\delta \int_M \alpha^n dV = 0$.

Theorem 7.1. *Let M be an n-dimensional closed oriented hypersurface of a euclidean $(n+1)$-space E^{n+1}. Then M is an S-hypersurface if and only if the function α and the scalar curvature r satisfy the equation*

$$\Delta\alpha^{n-1} + n(n-1)\alpha^{n+1} - \alpha^{n-1}r = 0. \tag{7.23}$$

Proof. If M is an S-hypersurface of E^{n+1}, then we have (7.22) for all differentiable functions φ. Hence we have (7.23). Conversely, if (7.23) holds, then we have, from (7.22), $\delta \int_M \alpha^n dV = 0$. Thus, M is an S-hypersurface. This proves the theorem. $\square$

Remark 7.1. Theorem 7.1 was proved by H. Hombu for $n = 2$.

Theorem 7.2. *If n is an odd integer > 1, then the only S-hypersurfaces in E^{n+1} are small hyperspheres.*

Proof. Let M be an S-hypersurface of a euclidean $(n+1)$-space E^{n+1}. By Problem II.2, we have

$$n(n-1)\alpha^2 \geqq r. \qquad (7.24)$$

Since n is an odd integer, α^{n+1} is nonnegative everywhere. Thus, by Theorem 7.1, we have

$$\Delta \alpha^{n-1} = \alpha^{n-1}\{r - n(n-1)\alpha^2\} \leqq 0. \qquad (7.25)$$

Hence, by Hopf's lemma, we see that either $\alpha = 0$ or $r = n(n-1)\alpha^2$. From Theorem 3.2 we know that the function α cannot be identically zero. Hence we have

$$n(n-1)\alpha^2 = r. \qquad (7.26)$$

From (7.26) we see that the subset $U = \{P \in M; \det(h_j^i) \neq 0 \text{ at } P\}$ is a nonempty subset of M and it is totally umbilical in E^{n+1}. Thus, from Proposition II.3.2, each component of U is contained in a small hypersphere of E^{n+1}; in particular, we see that $\det(h_j^i)$ is a nonzero constant on each component of U. This shows that U is also a closed subset of M. Hence, $M = U$. This proves the theorem. $\qquad \square$

Theorem 7.3. *If n is a positive even integer, then the only S-hypersurfaces in E^{n+1} such that the function α does not change its sign are small hyperspheres.*

Proof. Since α does not change its sign, (7.24) implies that

$$\alpha^{n-1}\{r - n(n-1)\alpha^2\} \qquad (7.27)$$

does not change its sign. Hence, if M is an S-hypersurface of E^{n+1}, then, by Theorem 7.1, we see that α^{n-1} is either a subharmonic function or a superharmonic function on M. Thus, by Hopf's lemma, we see that the function (7.27) vanishes. Hence, by a similar argument as the proof of Theorem 7.2, we can easily obtain the theorem. $\qquad \square$

Remark 7.2. Theorem 7.3 is due to H. Hombu for $n = 2$.

Remark 7.3. Willmore and Jhaveri (1972) have generalized Theorem 7.1 to a hypersurface in an arbitrary Riemannian manifold and used their result to study stable hypersurfaces in a Riemannian manifold with negative definite Ricci tensor.

Problems

1. Let M be a closed curve in a euclidean m-space E^m. Prove that the inequality (2.4) of Chern and Lashof is equal to the inequality (2.7) of Fenchel and Borsuk.

2. Prove Theorem 2.3.

3. Prove formula (2.2).

4. Prove formula (2.8).

5. Let $x: M \to E^m$ be an immersion of an n-dimensional closed manifold in E^m. Prove that (i) $TA_i(x) = 2$ for some i, $1 \leq i \leq n - 1$, when and only when M is imbedded as a small n-sphere in E^m. (ii) $\int_M |H|^n dV = c_n$ when and only when M is imbedded as a small n-sphere in E^m.

6. Let M be a closed surface in E^m. Prove that if the $(m-2)$-th curvature λ_{m-2} vanishes identically, then M is a convex hypersurface in a 3-dimensional linear subspace of E^m.

7. Let M be a complete oriented surface in E^m with $\lambda_1 = \lambda_2 = \cdots = \lambda_{m-2} = 0$. Prove that M is generated by a moving straight line with a fixed direction through a curve in E^m, that is, M is a cylinder in E^m (Shiohama, 1967).

8. Let M be a surface with nonpositive Gaussian curvature such that M is not homeomorphic to a torus. What is the value of

$$\inf(TA_1(x); x \text{ isometrical immersion of } M \text{ in } E^4\}?$$

9. Let $x: M \to E^m$ be an imbedding of an n-dimensional orientable closed manifold M in E^m. Prove that

   (i) The n-th mean curvature $G(\xi) \geq 0$ everywhere if and only if (a) M has no torsion, that is, $b(M) = b(M; \mathcal{R})$, (b) all odd-dimensional betti numbers of M vanish, that is, $b_i(M; \mathcal{R}) = 0$, $i = 1, 3, 5, \ldots$, and (c) $\tau(x) = b(M)$.

   (ii) If the n-th mean curvature $G(\xi) > 0$ everywhere, then the codimension is one, that is, $m = n + 1$, and M is imbedded as a convex hypersurface in E^{n+1}.

   If the codimension is one, the sufficiency of (i) was given in Chern and Lashof (1958).

10. Let M be an n-dimensional closed submanifold of a euclidean m-space E^m. Prove that, if for any unit normal vector ξ of M in E^m, the second fundamental tensor A_ξ is semidefinite, then M is contained in an $(n+1)$-dimensional linear subspace of E^m as a convex hypersurface. In particular, M is homeomorphic to a topological n-sphere (do Carmo and Lima, 1969).

Bibliography

Abe, K.
  1971. A characterization of totally geodesic submanifolds in S^N and CP^N by an inequality, Tohoku Math. J., *23*, 219–244.

Aeppli, A.
  1959. Einige Ähnlichkeits- und Symmetriesatze für differenzierbare Flächen in Raum, Comm. Math. Helv., *33*, 174–195.

Alexander, S.B.
  1969. Reducibility of euclidean immersions of low codimension, J. Differential Geometry, *3*, 69–82.

Alexandrov, A.D.
  1938. Ein allgemeiner Eindeutigkeitsatz für geschlossenen Flächen, C. R. Acad. Sci. URSS (Doklady), *19*, 227–229.
  1955. *Die innere Geometrie der Konvexen Flächen*, Akademie–Verlag, Berlin.
  1956. Uniaueness theorems for surfaces in the large, I. Vestnik Leningrad Univ., *11*, No. 19, 5–7.
  1962. A characteristic property of spheres, Ann. Math. Pura Appl., *58*, 303–315.

Allendoerfer, C.B.
  1940. The Euler number of a Riemannian manifold, Amer. J. Math., *62*, 243–248.

Allendoerfer, C.B. and Weil, A.
    1943.   The Gauss-Bonnet theorem for Riemannian polyhedra, Trans. Amer. Math. Soc., *53*, 101–129.

Almgren, F.J. Jr.
    1966.   Some interior regularity theorems for minimal surfaces and extension of Bertein's theorem, Ann. of Math., *84*, 277–292.

Ambrose, W.
    1960.   The Cartan structural equations in classical Riemannian geometry, J. Indian Math. Soc., *24*, 23–76.

Auslander, L. and Mackenzie, R.
    1963.   *Introduction to Differentiable Manifolds*, McGraw–Hill, New York.

Avez, A.
    1970.   Characteristic classes and Weyl tensor; applications to general relativity, Proc. Nat. Acad. Sci. USA, *66*, 265–268.

Banchoff, T.F.
    1965.   Tightly embedded 2-dimensional polyhedra manifolds, Amer. J. Math., *87*, 462–472.
    1967.   Critical points and curvature for embedded polyhedra, J. Differential Geometry, *1*, 245–256.
    1970.   The spherical two-piece property and tight surfaces in spheres, J. Differential Geometry, *4*, 193–205.
    1971.   The two-piece property and tight n-manifolds with boundary in E^n, Trans. Amer. Math. Soc., *161*, 259–267.

Bishop, R.L.
    1968.   The holonomy algebra of immersed manifolds of codimension two, J. Differential Geometry, *2*, 347–253.

Bishop, R.L. and Crittenden, R.J.
    1964.   *Geometry of Manifolds*, Academic, New York.

Bishop, R.L. and Goldberg, S.I.
    1964.   Some implications of the generalized Gauss-Bonnet theorem, Trans. Amer. Math. Soc., *112*, 508–535.

Bishop, R.L. and O'Neill, B.
    1969.   Manifolds of negative curvature, Trans. Amer. Math. Soc., *145*, 1–49.

Blair, D.E. and Stone, A.P.
    1970.   Geometry of manifolds which admit conservation laws, Ann. Inst. Fourier, Grenoble, *21*, 1–9.
    1971.   Geometry of manifolds which admit conservation laws, II, Math. Ann., *194*, 167–179.

Bombieri, E., de Giorgi, and Giusti, E.
    1969.   Minimal cones and the Bernstein problem, Invent. Math., *7*, 243–268.

Borsuk, K.
    1947.   Sur la courbure totale des courbes, Ann. de la Soc. Polonaise, *20*, 251–265.

Bott, R.
    1970.   On a topological obstruction to integrability, Proc. Symp. in Pure Math., American Mathematical Society, *16*, 127–131.

Braidi, S. and Hsiung, C.C.
    1970.   Submanifolds of spheres, Math. Z., *115*, 235–251.

Calabi, E.
    1953a.   Isometric imbedding of complex manifolds, Ann. of Math., *58*, 1–23.
    1953b.   *Metric Riemann surfaces, Contributions to the Theory of Riemann Surfaces*, Princeton Univ. Press, Princeton, New Jersey, 77–85.
    1967.   Minimal immersions of surfaces in euclidean spheres, J. Differential Geometry, *1*, 111–126.

do Carmo, M.P. and Lima, E.
    1969.   Isometric immersions with semi-definite quadratic forms, Arch. der Math. *20*, 173–175.

do Carmo, M.P. and Wallach, N.R.
    1970.   Representations of compact groups and minimal immersions into spheres, J. Differential Geometry, *4*, 91–104.
    1971.   Minimal immersions of spheres into spheres, Ann. of Math., *93*, 43–62.

Cartan, É.
    1917.   La déformation des hypersurfaces dans l'espace conforme reel a $n \geq 5$ dimensions, Bull. Soc. Math. France, *45*, 57–121.
    1919.   Sur les variétés de courbure constante d'un espace euclidien ou non euclidien, Bull. Soc. Math. France, *47* (1919), 125–160; *48* (1920), 132–208.
    1946.   *Lecons sur la Géométrie des Espaces de Riemann*, Gauthier-Villars, Paris.

Chen, B.-Y.
    1967.   On the total absolute curvature of manifolds immersed in Riemannian manifolds, Kōdai Math. Sem. Rep., *19*, 299–311.
    1968.   Some integral formulas of the Gauss-Kronecker curvature, Kōdai Math. Sem. Rep., *20*, 410–413.
    1969.   Surfaces of curvature $\lambda_N = 0$ in E^{2+N}, Kōdai Math. Sem. Rep., *21*, 331–334.

1970a. On an inequality of T.J. Willmore, Proc. Amer. Math. Soc., *26*, 473–479; Errata, *30* (1971), 627.

1970b. On the total absolute curvature of manifolds immersed in Riemannian manifolds, II, Kōdai Math. Sem. Rep., *22*, 98–106.

1971a. On an inequality of mean curvature of higher degree, Bull. Amer. Math. Soc., *77*, 157–159.

1971b. On the total curvature of immersed manifolds, I: An inequality of Fenchel-Borsuk-Willmore, Amer. J. Math., *93*, 148–162.

1971c. Submanifolds in a euclidean hypersphere, Proc. Amer. Math. Soc., *27*, 627–628.

1971d. Some results of Chern-do Carmo-Kobayashi type and the length of second fundamental form, Indiana Univ. Math. J., *20*, 1175–1185; Errata, *22* (1972), 399.

1971e. Minimal hypersurfaces of an m-sphere, Proc. Amer. Math. Soc., *29*, 375–380.

1971f. Minimal submanifolds in a euclidean hypersphere, Yokohama Math. J., *19*, 45–47.

1971g. On a theorem of Fenchel-Borsuk-Willmore-Chern-Lashof, Math. Ann., *194*, 19–26.

1971h. Some integral formulas for hypersurfaces in euclidean space, Nagoya Math. J., *43*, 117–125.

1971i. On the mean curvature of submanifolds of euclidean space, Bull. Amer. Math. Soc., *77*, 741–743.

1971j. On the scalar curvature of immersed manifolds, Math. J. Okayama Univ., *15*, 7–14.

1971k. On the support functions and spherical submanifolds, Math. J. Okayama Univ., *15*, 15–23.

1971l. On an integral formula of Gauss-Bonnet-Grotemeyer, Proc. Amer. Math. Soc., *28*, 208–212.

1971m. "Notes on Characteristic Classes", Mimeographic Notes, Michigan State University, East Lansing, Michigan.

1972a. On an inequality of mean curvature, J. London Math. Soc., *4*, 647–650.

1972b. On the total curvature of immersed manifolds, II: Mean curvature and length of second fundamental form, Amer . J. Math., *94*, 899–909.

1972c. Minimal surfaces of S^m with Gauss curvature ≤ 0, Proc. Amer. Math. Soc., *31*, 235–238.

1972d. Minimal surfaces with constant Gauss curvature, Proc. Amer. Math. Soc., *34*, 504–508.

1972e. Surfaces with parallel mean curvature vector, Bull. Amer. Math. Soc., *78*, 709–710.

1972f. On a variational problem of hypersurfaces, J. London Math. Soc., *6*, 321–325.

1972g. G-total curvature of immersed manifolds, J. Differential Geometry, *7*, 373–393.

1973a. Pseudoumbilical surfaces with constant Gauss curvature, Proc. Edinburg Math. Soc., *18*, 143–148.

1973b. An invariant of conformal mappings, Proc. Amer. Math. Soc., *40*, 563–564.

1973c. On the surfaces with parallel mean curvature vector, Indiana Univ. Math. J., *22*, 655–666.

1973d. Pseudoumbilical submanifolds in a Riemannian manifold of constant curvature, II, J. Math. Soc. Japan, *25*, 105–114.

1973e. A characterization of standard flat tori, Proc. Amer. Math. Soc., *37*, 564–567.

1973f. On the total curvature of immersed manifolds, III: Surfaces in euclidean 4-space, Amer. J. Math., *95*, 636–642.

Chen, B.-Y. and Houh, C.-S.
  1972. Some differential geometric inequalities for surfaces in euclidean spaces, Tensor, *23*, 105–109.

Chen, B.-Y. and Ludden, G.D.
  1972a. Rigidity theorems for surfaces in euclidean space, Bull. Amer. Math. Soc., *78*, 72–73; Errata, *78* (1972), 628.

  1972b. Surfaces with mean curvature vector parallel in the normal bundle, Nagoya Math. J., *47*, 161–168.

Chen, B.-Y. and Ogiue, K.
  1973. On the scalar curvature and sectional curvatures of a Kaehler submanifold, Proc. Amer. Math. Soc., *41*, 247–250.

  1974. On totally real submanifolds, Trans. Amer. Math. Soc., *193*, 257–266.

Chen, B.-Y. and Okumura, M.
  1973. Scalar curvature, inequality and submanifold, Proc. Amer. Math. Soc., *38*, 605–608.

Chen, B.-Y. and Yano, K.
  1971a. Some integral formulas for submanifolds and their applications, J. Differential Geometry, *5*, 467–477.

  1971b. On submanifolds of submanifolds of a Riemannian manifold, J. Math. Soc. Japan, *23*, 548–554.

  1972a. Pseudoumbilical submanifolds in a Riemannian manifold of constant curvature, Differential Geometry in honor of K. Yano, Kinokuniya, Tokyo, 61–71.

  1972b. Sous-variété localement conformes à un espace euclidean, C.R. Acad. Sci. Paris, *275*, 123–126.

  1972c. Special quasiumbilical hypersurfaces and locus of spheres, Atti. Acad. Naz. Lincei, Rend., *53*, 255–260.

  1972d. Hypersurfaces of a conformally flat space, Tensor, *26*, 318–322.

1972e. "Conformally flat submanifolds," Mimeographic Notes, Michigan State University.

1973a. Special conformally flat spaces and canal hypersurfaces, Tôhoku Math. J., 25, 177–184.

1973b. Conformally flat spaces of codimension 2 in a euclidean space, Canadian J. Math., 25, 1170–1173.

1973c. Umbilical submanifold with respect to a nonparallel normal direction, J. Differential Geometry, 8, 589–597.

1973d. Submanifolds umbilical with respect to a nonparallel normal subbundle, Kōdai Math. Sem. Rep., 25, 289–296.

1973e. Pseudoumbilical submanifolds of codimension 3 with constant mean curvature, Kōdai Math. Sem. Rep., 25, 490–501.

1973f. Submanifolds umbilical with respect to a quasiparallel normal direction, Tensor, 27, 41–44.

Chern, S.S.

1944a. A simple intrinsic proof of the Gauss-Bonnet formula for closed Riemannian manifolds, Ann. of Math., 45, 747–752.

1944b. On a theorem of algebra and its geometrical applications, J. Indian Math. Soc., 8, 29–36.

1945a. On the curvature integral in a Riemannian manifold, Ann. of Math., 46, 674–684.

1945b. Some new characterizations of euclidean sphere, Duke Math. J., 12, 279–290.

1946a. Characteristic classes of Hermitian manifolds, Ann. of Math., 47, 85–121.

1946b. Some new viewpoints in differential geometry in the large, Bull. Amer. Math. Soc., 52, 1–30.

1951a. Topics in Differential Geometry, Institute for Advanced Study, Princeton.

1951b. La géométrie des sous-variétés d'un espace euclidien à plusieurs dimensions, Enseignement Math., 40 (1951–1954), 26–46.

1953a. Relations between Riemannian and Hermitian geometries, Duke Math. J. 20, 575–587.

1953b. Some formulas in theory of surfaces, Bol. Soc. Mat. Mexicana, 10, 30–40.

1955a. On curvature and characteristic classes of a Riemannian manifold, Abh. Math. Sem. Univ. Hamburg, 20, 117–126.

1955b. An elementary proof of the existence of isothermal parameters on a surface, Proc. Amer. Math. Soc., 6, 771–782.

1957. A proof of the uniqueness of Minkowski's problem for convex surfaces, Amer. J. Math., 79, 949–950.

1958. Geometry of submanifolds in a complex projective space, Intern. Symp. Algebraic Topology, Mexico City, 87–96.

1959a. "Differentiable Manifolds," University of Chicago Notes.

1959b. "Complex Manifolds," University of Chicago Notes.

1959c. Integral formulas for hypersurfaces in euclidean space and their applications to uniqueness theorem, J. Math. Mech. *8*, 947–955.

1965a. *Minimal surfaces in an euclidean space of n dimensions, Symp. Differential and Combinatorial Topology in honor of M. Morse*, Princeton Univ. Press, Princeton, New Jersey, 187–198.

1965b. On the curvature of a piece of hypersurface in euclidean space, Abh. Math. Sem. Univ. Hamburg, *29*, 77–91.

1965c. On the differential geometry of a piece of submanifold in euclidean space, Proc. US-Japan Seminar in Differential Geometry, Kyoto, 17–21.

1965d. *Lectures on Integral Geometry*, Academic Sinica, Taipei.

1966. The geometry of G-structures, Bull. Amer. Math. Soc., *72*, 167–219.

1967a. On Einstein hypersurfaces in a Kaehlerian manifold of constant holomorphic sectional curvature, J. Differential Geometry, *1*, 21–31.

1967b. Curves and surfaces in euclidean spaces, Studies in Global Geometry and Analysis, Math. Assoc. Amer., 16–56.

1968. "Minimal Submanifolds in a Riemannian Manifolds," University of Kansas, Technical Report 19.

1969. Simple proofs of two theorems on minimal surfaces, Enseignement Math., *15*, 53–61.

1970a. Holomorphic curve and minimal surfaces, Carolina Conference Proceedings, University of North Carolina, 1–28.

1970b. On minimal spheres in the four-sphere, Studies and Essays presented to Yu-Why Chen on his Sixtieth Birthday, 137–150.

1970c. Differential geometry; its past and its future, Actes Congrès Intern. Math., *1*, 41–53.

Chern, S.S., do Carmo, M., and Kobayashi, S.

1970. Minimal submanifolds of a sphere with second fundamental form of constant length, *Functional Analysis and Related Fields*, Springer-Verlag, 59–75.

Chern, S.S., Hano, J., and Hsiung, C.C.

1960. A uniqueness theorem on closed convex hypersurfaces, J. Math. Mech., *9*, 85–88.

Chern, S.S. and Hsiung, C.C.

1962/63. On the isometry of compact submanifolds in euclidean space, Math. Ann., *149*, 278–285.

Chern, S.S. and Kuiper, N.

1952. Some theorems on the isometric imbedding of compact Riemann manifolds in euclidean space, Ann. of Math., *56*, 422–430.

Chern, S.S. and Lashof, R.K.
    1957. On the total curvature of immersed manifolds, Amer. J. Math., *79*, 306–318.
    1958. On the total curvature of immersed manifolds, II, Michigan Math. J., *5*, 5–12.

Chern, S.S. and Simons, J.
    1971. Some cohomology classes in principal fibre bundles and their application to Riemannian geometry, Proc. Nat, Acad. Sci., USA, *68*, 791–794.
    1974. Characteristic forms and geometric invariants, Ann. of Math., *99*, 48–69.

Chevalley, C.
    1946. *Theory of Lie Groups*, Princeton Univ. Press, Princeton, New Jersey.

Courant, R. and Hilbert, D.
    1962. *Methods of Mathematical Physics*, Wiley-Interscience, New York.

Davies, E.T.J.
    1942. The first and second variations of the volume integral in Riemannian space, Quart. J. Math. Oxford, *13*, 58–64.

Eells, J. Jr.
    1959. A generalization of the Gauss-Bonnet theorem, Trans. Amer. Math. Soc., *92*, 142–153.
    1966. A setting for global analysis, Bull. Amer. Math. Soc., *72*, 751–807.

Eells, J. Jr. and Kuiper, N.H.
    1957. Manifolds which are like projective planes, I.H.E.S. Publ. Math., *14*, 181–222.

Eells, J. Jr. and Sampson, H.
    1964. Harmonic mappings of Riemannian manifolds, Amer. J. Math., *86*, 109–160.
    1965. Variational theory in fibre bundles, Proc. US-Japan Seminar in Differential Geometry, Kyoto, 22–33.

Eisenhart, L.P.
    1927. *Non-Riemannian Geometry*, Amer. Math. Soc. Colloq. Publ. 8.
    1947. *An Introduction to Differential Geometry*, Princeton Univ. Press, Princeton, New Jersey.
    1949. *Riemannian Geometry*, Princeton Univ. Press, Princeton, New Jersey.

Erbacher, J.A.
    1971. Reduction of the codimension of an isometric immersion, J. Differential Geometry, *5*, 333–340.
    1972a. Isometric immersions with constant mean curvature and triviality of the normal bundle, Nagoya Math. J., *45*, 139–165.

1972b.   A variational problem for submanifolds of euclidean space, *36*,
         467–470.

Fary, I.
1949.   Sur la courbure totale d'une courbe gauche faisant un noeud, Bull.
        Soc. Math. France, *77*, 128–138.

Feldman, E.A.
1965.   The geometry of immersions, I, Trans. Amer. Math. Soc., *120*,
        185–224.
1966.   Geometry of immersions, II, Trans. Amer. Math. Soc., *125*, 181–215.
1967.   On parabolic and umbilical points of immersed hypersurfaces, Trans.
        Amer. Math. Soc., *127*, 1–28.

Fenchel, W.
1929.   Über die Krümmung und windüng geschlossener Raumkurven, Math.
        Ann., *101*, 238–252.
1940.   On total curvatures of Riemannian manifolds, J. London Math. Soc.,
        15, 15–22.

Ferus, D.
1967.   Über die absolute Totalkrümmung höher-dimensionaler Knoten,
        Math. Ann., *171*, 81–86.
1968.   *Totale absolutkrummung in Differential-geometrie und-topologie*,
        Lecture Notes in Mathematics, *66*, Springer-Verlag, Berlin.
1970.   On the type number of hypersurfaces in spaces of constant curvature,
        Math. Ann., *187*, 310–316.
1971a. On the completeness of nullity foliations, Michigan Math. J., *18*,
        61–64.
1971b. The torsion form of submanifolds in E^N, Math. Ann., *193*, 114–120.

Fialkow, A.
1938.   Hypersurfaces of a space of constant curvature, Ann. of Math., *39*,
        762–785.

Flanders, H.
1963.   *Differential Forms*, Academic, New York.
1966.   Remark on mean curvature, J. London Math. Soc., *41*, 364–366.

Frankel, T.T.
1966.   On the fundamental group of a compact minimal submanifold, Ann.
        of Math., *83*, 68–73.

Friedman, A.
1961.   Local isometric imbeddings of Riemannian manifolds with indefinite
        metrics, J. Math. Mech., *10*, 625–649.
1965.   Isometric embedding of Riemannian manifolds into euclidean spaces,
        Rev. Mod. Phys., *37*, 201–203.

Gardner, R.B.
    1969.   An integral formula for immersions in euclidean space, J. Differential Geometry, *3*, 245–252.
    1970.   The Dirichlet integral in differential geometry, Proc. Symp. Pure Math., American Mathematical Society, *15*, 231–237.
    1972.   Subscalar pairs of metrics and hypersurfaces with nondegenerate second fundamental form, J. Differential Geometry, *6*, 437–458.

de Giorgi, E.
    1965.   Una estensione del teorema di Bernstein, Ann. Della Scuola Normale Superiore di Piza, Scienze Fis. Mat. III, XIX, *1*, 79–85.

Goldberg, S.I.
    1962.   *Curvature and Homology*, Academic, New York.
    1969.   On conformally flat spaces with definite Ricci curvature, Kōdai Math. Sem. Rep., *21*, 226–232.

Goldberg, S.I. and Kobayashi, S.
    1967.   Holomorphic bisectional curvature, J. Differential Geometry, *1*, 225–233.

Gray, A.
    1965.   Minimal varieties and almost Hermitian submanifolds, Michigan Math. J., *12*, 273–287.
    1969.   Vector cross products on manifolds, Trans. Amer. Math. Soc., *141*, 465–504.

Green, L.W.
    1954.   Surfaces without conjugate points, Trans. Amer. Math. Soc., *76*, 529–546.
    1960.   A sphere characterization related to Blaschke's conjecture, Pacific J. Math., *10*, 837–841.

Greene, R.E.
    1970.   Isometric embeddings of Riemannian and pseudo-Riemannian manifolds, Memoirs Amer. Math. Soc. No. 97.
    1971.   Metrics and isometric embeddings of the 2-sphere, J. Differential Geometry, *5*, 353–356.

Greene, R.E. and Wu, H.
    1971.   On the rigidity of punctured ovaloids, Ann. of Math., *94*, 1–20.

Grossman, N.
    1972.   Relative Chern-Lashof theorems, J. Differential Geometry, *7*, 607–614.

Guggenheimer, H.
    1963.   *Differential Geometry*, McGraw-Hill, New York.

Hardy, G.H., Littlewood, J.E., and Polya, G.
  1934. *Inequalities*, Cambridge Univ. Press, London and New York.

Harle, C.E.
  1971. Rigidity of hypersurfaces of constant scalar curvature, J. Differential
          Geometry, *5*, 85–112.

Hartman, P.
  1965. On isometric immersions in euclidean space of manifolds with
          nonnegative sectional curvatures, Trans. Amer. Math. Soc., *115*,
          94–109.
  1970. On the isometric immersions in euclidean space of manifolds with
          nonnegative sectional curvatures, II, Trans. Amer. Math. Soc., *147*,
          529–540.

Hartman, P. and Nirenberg, L.
  1959. On spherical image maps whose Jacobians do not change sign, Amer.
          J. Math., *81*, 901–920.

Hartman, P. and Winter, A.
  1950a. On the embedding problem in differential geometry, Amer. J. Math.,
           72, 553–564.
  1950b. The fundamental equations of differential geometry, Amer. J. Math.,
           72, 757–774.

Helgason, S.
  1962. *Differential Geometry and Symmetric Spaces*, Academic, New York.

Hermann, R.
  1964. Spherical compact hypersurfaces, J. Math. Mech., *13*, 237–242.
  1965. Vanishing theorems for homology of submanifolds, J. Math. Mech.,
          14, 479–483.

Hicks, N.
  1963a. Submanifolds of semi-Riemannian manifolds, Rend. Circ. Mat.
           Palermo (2), *12*, 137–149.
  1963b. *Notes on Differential Geometry*, Van Nostrand, Princeton, New Jersey.

Hirzebruch, F.
  1966. *Topological Methods in Algebraic Geometry*, Springer-Verlag,
          New York.

Hoffman, D.A.
  1972. Surfaces in constant curvature manifolds with parallel mean curvature
          vector field, Bull. Amer. Math. Soc., *78*, 247–250.
  1973. Surfaces of constant mean curvature in constant curvature manifold,
          J. Differential Geometry, *8*, 161–176.

Hopf, H.
  1951.   Über Flächen mit einer Relation zwischen den Hauptkrümmungen, Math. Nachr., *4*, 232–249.

Hopf, H. and Rinow, W.
  1931.   Über den Begriff der vollständigen differentialgeometrische Fläche, Comm. Math. Helv., *3*, 209–225.

Hopf, H. and Voss, K.
  1952.   Ein Satz aus der Flächentheorie im Grossen, Arch. Math., *3*, 187–192.

Houh, C.S.
  1972a. On an integral formula for closed hypersurfaces of the sphere, Proc. Amer. Math. Soc., *35*, 234–237.
  1972b. Surfaces with maximal Lipschitz-Killing curvature in the direction of mean curvature vector, Proc. Amer. Math. Soc., *35*, 537–542.
  1972c. Some surfaces in a space of constant curvature, Special Issue Dedicated to Puh Pan, The Chinese University of Hong Kong, 153–155.

Hsiang, W.-Y.
  1966.   On the compact homogeneous minimal submanifolds, Proc. Nat. Acad. Sci. USA, *56*, 5–6.
  1967.   Remarks on closed minimal submanifolds in the standard Riemannian m-sphere, J. Differential Geometry, *1*, 257–267.

Hsiung, C.C.
  1954.   Some integral formulas for hypersurfaces, Math. Scand. *2*, 286–294.
  1956.   On differential geometry of hypersurfaces in the large, Trans. Amer. Math. Soc., *81*, 243–252.
  1957.   Some global theorems on hypersurfaces, Canad. J. Math., *9*, 5–14.
  1958.   A uniqueness theorem for Minkowski's problem for convex surfaces with boundary, Illinois J. Math., *2*, 71–75.
  1961.   Isoperimetric inequalities for 2-dimensional Riemannian manifolds with boundary, Ann. of Math., *73*, 213–220.

Hsiung, C.C. and Rhodes, B.H.
  1968.   Isometrics of compact submanifolds of a Riemannian manifold, J. Differential Geometry, *2*, 9–24.

Huber, A.
  1957.   On subharmonic functions and differential geometry in the large, Comm. Math. Helv., *32*, 13–72.

Husemoller, D.
  1966.   *Fibre Bundles*, McGraw-Hill, New York.

Itoh, T.
  1970.   Complete surfaces in E^4 with constant mean curvature, Kōdai Math. Sem. Rep., *22*, 150–158.

1973. Minimal surfaces in a Riemannian manifold of constant curvature, Kōdai Math. Sem. Rep., *25*, 202–214.

Jacobson, N.
1962. *Lie Algebra*, Wiley-Interscience, New York.

Katsurada, Y.
1962. Generalized Minkowski formulas for closed hypersurfaces in Riemannian space, Ann. Math. Pura Appl., *57*, 283–293.

Kenmotsu, K.
1970. Some remarks on minimal submanifolds, Tôhoku Math. J., *22*, 240–248.

Klingenberg, W.
1959. Contributions to Riemannian geometry in the large, Ann. of Math., *69*, 654–666.

Klotz, T. and Osserman, R.
1966/67. Complete surfaces in E^3 with constant mean curvature, Comm. Math. Helv., *41*, 313–318.

Kobayashi, S.
1956. Induced connections and imbedded Riemannian spaces, Nagoya Math. J., *10*, 15–25.
1958. Compact homogeneous hypersurfaces, Trans. Amer. Math. Soc., *88*, 137–143.
1967a. Imbeddings of homogeneous space with minimum total curvature. Tôhoku Math. J., *19*, 63–74.
1967b. Hypersurfaces of complex projective space with constant scalar curvature. J. Differential Geometry, *1*, 369–370.
1968. Isometric imbeddings of compact symmetric spaces, Tôhoku Math. J., *20*, 21–25.
1971. *Hyperbolic Manifolds and Holomorphic Mappings*, Marcel Dekker, New York.

Kobayashi, S. and Eells, J. Jr.
1965. Problems in differential geometry, Proc. US-Japan Seminar in Differential Geometry, Kyoto, 167–177.

Kobayashi, S. and Nomizu, K.
1963. *Foundations of Differential Geometry, I, II*, Wiley-Interscience, New York.

Kuiper, N.H.
1949. On conformally flat spaces in the large, Ann. of Math., *50*, 916–924.
1950. On compact conformally euclidean spaces of dimension > 2, Ann. of Math., *52*, 478–490.

1955. On C^1-isometric imbeddings I, Indag. Math., *17*, 545–556; II, 683–689.

1958. Immersions with minimal absolute curvature, Colloque de Geometric Differentielle Globale, Bruxelles, 75–87.

1960. On surfaces in euclidean three-space, Bull. Soc. Math. Belg., *12*, 5–22.

1961. Convex immersions of closed surfaces in E^3. Nonorientable closed surfaces in E^3 with minimal total absolute Gauss-curvature, Comm. Math. Helv., *35*, 85–92.

1967. Der Satz von Gauss-Bonnet für Abbildunger in E^N und damit verwandte Probleme, Jber. Deut. Math. Ver., *69*, 77–88.

1970. Minimal total absolute curvature for immersions, Inventions Math., *10*, 209–238.

Kulkarni, R.S.

1967. Curvature and metric, Thesis, Harvard Univ.

1972. Conformally flat manifolds, Nat. Acad. Sci. USA, *69*, 2675–2676.

Lawson, H.B. Jr.

1968. Minimal varieties in constant curvature manifolds, Thesis, Stanford University.

1969. Local rigidity theorems for minimal hypersurfaces, Ann. of Math., *89*, 187–197.

1970a. The global behavior of minimal surfaces in S^n, Ann. of Math., *92*, 224–237.

1970b. Complete minimal surfaces in S^3, Ann. of Math., *92*, 335–374.

1970c. The unknottedness of minimal embeddings, Inventions Math., *11*, 183–187.

1970d. Rigidity theorems in rank-1 symmetric spaces, J. Differential Geometry, *4*, 349–357.

1970e. Compact minimal surfaces in S^3, Proc. Symp. in Pure Math., *15*, American Mathematical Society, 275–282.

1971. Some intrinsic characterizations of minimal surfaces, J. D'Analyse Math., *24*, 151–161.

Leung, D.S. and Nomizu, K.

1971. The axiom of spheres in Riemannian geometry, J. Differential Geometry, *5*, 487–489.

Lichnerowicz, A.

1955. *Théorie Globale des Connexions et des Groupes d'Holonomie*, Ed. Cremonese, Rome.

1958. *Géométrie des Groupes de Transformations*, Dunod, Paris.

Liebmann, H.

1901. Über Flächen von konstanter Gaußscher Krümmung, Trans. Amer. Math. Soc., *2*, 87–99.

Little, J.A.
   1969. On singularities of submanifolds of higher dimensional euclidean spaces, Ann. di Math. (IV), *83*, 261–336.
   1970. Nondegenerate homotopies of curves on the unit 2-sphere, J. Differential Geometry, *4*, 339–348.
   1971a. Nondegenerate homotopies of spherical curves, Differentialgeometrie in Grossen, Oberwolfach, 241–249.
   1971b. Third order nondegenerate homotopies of space curves, J. Differential Geometry, *5*, 503–515.

Little, J.A. and Pohl, W.
   1971. Smooth tight embeddings of high codimension, Inventions Math., *13*, 179–204.

Loos, O.
   1969. *Symmetric Spaces, I, II*, Benjamin, New York.

Maltz, R.
   1966. The nullity spaces of the curvature operator, Cahiers de Topologie et Geometrie Differentielle, *8*, 1–20.
   1971. Isometric immersions into spaces of constant curvature, Illinois J. Math., *15*, 490–502.

Massey, W.S.
   1962. Surfaces of Gaussian curvature zero in euclidean 3-space, Tôhoku Math. J., *14*, 73–79.

Matsushima, Y.
   1971. Vector bundle valued harmonic forms and immersions of Riemannian manifolds, Osaka J. Math., *8*, 1–13.

Milnor, J.W.
   1956. On manifolds homeomorphic to the 7-sphere, Ann. of Math., *64*, 394–405.
   1958. "Lectures on Characteristic Classes," Notes from Princeton Univ.
   1963. *Morse Theory*, Ann. Math. Studies, No. 51, Princeton Univ. Press, Princeton, New Jersey.
   1965. *Topology from the Differentiable viewpoint*, Univ. Virginia Press.

Moore, J.D.
   1971. Isometric immersions of Riemannian products, J. Differential Geometry, *5*, 159–168.

Morse, M.
  1934. *The Calculus of Variations in the Large*, American Mathematical
          Society, Providence, Rhode Island.

Myers, S.B.
  1941. Riemannian manifolds with positive mean curvature, Duke Math. J.,
          8, 401–404.
  1951. Curvature of closed hypersurfaces and nonexistence of closed minimal
          hypersurfaces, Trans. Amer. Math. Soc., *71*, 211–217.

Nagano, T.
  1958. Sur des hypersurfaces et quelques groups d'isométries d'un espace
          Riemannien, Tôhoku Math. J., *10*, 242–252.
  1959. The conformal transformation on a space with parallel Ricci tensor,
          J. Math. Soc. Japan, *11*, 10–14.
  1967. A problem on the existence of an Einstein metric, J. Math. Soc. Japan,
          19, 30–31.
  1969. Homotopy invariants in differential geometry, I, Trans. Amer. Math.
          Soc., *144*, 441–455.
  1970. *Homotopy Invariants in Differential Geometry*, Memoirs Amer. Math.
          Soc., *100*, American Mathematical Society, Providence, Rhode Island.

Nagano, T. and Takahashi, T.
  1960. Homogeneous hypersurfaces in euclidean spaces, J. Math. Soc. Japan,
          12, 1–7.

Nash, J.F.
  1954. C^1-isometric imbeddings, Ann. of Math., *60*, 383–396.
  1956. The imbedding problem for Riemannian manifolds, Ann. of Math.,
          63, 20–63.

Nirenberg, L.
  1953. The Weyl and Mirikowski problems in the differential geometry in the
          large, Comm. Pure Appl. Math., *6*, 337–394.

Nitsche, J.C.C.
  1965. On recent results in the theory of minimal surfaces, Bull. Amer. Math.
          Soc., *71*, 195–270.

Nomizu, K.
  1956. *Lie Groups and Differential Geometry*, Publ. Math. Soc. Japan No. 2.
  1968. On hypersurfaces satisfying a certain condition on the curvature tensor,
          Tôhoku Math. J., *20*, 46–59.
  1973. Generalized central spheres and the notion of spheres in Riemannian
          geometry, Tôhoku Math. J., *25*, 129–137.

Nomizu, K. and Smyth, B.
  1968. Differential geometry of complex hypersurfaces, II, J. Math. Soc.
          Japan, *20*, 498–521.

1969a.   A formula of Simon's type and hypersurfaces with constant mean curvature, J. Differential Geometry, *3*, 367–377.

1969b.   On the Gauss mapping for hypersurfaces of constant mean curvature in the sphere, Comm. Math. Helv., *44*, 484–490.

Obata, M.

1971.   The conjectures on conformal transformations of Riemannian manifolds, J. Differential Geometry, *6*, 247–258.

Ogiue, K.

1969a.   Complex submanifolds of the complex projective space with second fundamental form of constant length, Kōdai Math. Sem. Rep., *21*, 252–254.

1969b.   Complex hypersurfaces of a complex projective space, J. Differential Geometry, *3*, 253–256.

1971.   Scalar curvature of complex submanifolds of a complex projective space, J. Differential Geometry, *5*, 229–232.

1972a.   Positively curved complex hypersurfaces immersed in a complex projective space, Tôhoku Math. J., *24*, 51–54.

1972b.   Differential Geometry of algebraic manifolds, Differential Geometry, in honor of K. Yano, Kinokuniya, Tokyo, 355–372.

1974.   Differential Geometry of Kaehler Submanifolds, Adv. Math., *13*, 73–114.

O'Neill, B.

1959.   An algebraic criterion for immersion, Pacific J. Math., *9*, 1239–1247.

1960.   Immersion of manifolds of nonpositive curvature, Proc. Amer. Math. Soc., *11*, 132–134.

1962a.   Isometric immersion of flat Riemannian manifolds in euclidean space, Michigan Math. J., *9*, 199–205.

1962b.   Isometric immersions which preserve curvature operator, Proc. Amer. Math. Soc., *13*, 759–763.

O'Neill, B. and Stiel, E.

1963.   Isometric immersions of constant curvature manifolds, Michigan Math. J., *10*, 335–339.

Osserman, R.

1960.   On the Gauss curvature of minimal surfaces, Trans. Amer. Math. Soc., *96*, 115–128.

1964.   Global properties of minimal surfaces in E^3 and E^N, Ann. of Math., *80*, 340–364.

1965.   Global properties of classical minimal surfaces, Duke Math. J., *32*, 565–573.

1969a.   Minimal varieties, Bull. Amer. Math. Soc., *75*, 1092–1120.

1969b.   *A Survey of Minimal Surfaces*, Van Nostrand, Princeton, New Jersey.

Ōtsuki, T.
  1953.   On the existence of solutions of a system of quadratic equations and its geometrical application, Proc. Japan Acad., *29*, 99–100.

  1954.   Isometric imbedding of Riemannian manifolds in a Riemannian manifold, J. Math. Soc. Japan, *6*, 221–234.

  1956.   Note on the isometric imbedding of compact Riemannian manifolds in euclidean spaces, Math. J. Okayama Univ., *5*, 95–102.

  1966.   On the total curvature of surfaces in euclidean spaces, Japan J. Math., *35*, 61–71.

  1968a. A theory of Riemannian submanifolds, Kōdai Math. Sem. Rep., *20*, 282–295.

  1968b. Pseudo-umbilical submanifolds with m-index ≤ 1 in euclidean spaces, Kōdai Math. Sem. Rep., *20*, 296–304.

  1970.   Minimal hypersurfaces in a Riemannian manifold of constant curvature, Amer. J. Math., *92*, 145–173.

  1971a. Minimal submanifolds with m-index 2, J. Differential Geometry, *6*, 193–211.

  1971b. Minimal submanifolds with m-index 2 in Riemannian manifolds of constant curvature, Tôhoku Math. J., *23*, 371–402.

  1972a. On internal inequalities related with a certain nonlinear differential equation, Proc. Japan Acad., *48*, 9–12.

  1972b. On a 2-dimensional Riemannian manifold, Differential Geometry in honor of K. Yano, Kinokuniya, Tokyo, 401–414.

  1972c. Minimal submanifolds with m-index 2 and generalized Veronese surfaces, J. Math. Soc. Japan, *24*, 89–122.

Pinl, M. and Trapp, H.
  1968.   Stationäre Krümmungsdichten auf Hyperflächen dès euklideschen R_{n+1}, Math. Ann., *176*, 257–292.

Reeb, G.
  1952.   Sur certaines propriétés topologiques des variétés feuilletées, Actual. Sci. et Indus., *1183*, 91–154.

Reilly, R.C.
  1970.   Extrinsic rigidity theorems for compact submanifolds of the sphere, J. Differential Geometry, *4*, 487–497.

  1973.   Variational properties of functions of the mean curvatures for hypersurfaces in space forms, J. Differential Geometry, *8*, 465–477.

de Rham, G.
  1955.   *Variétés Différentiables*, Hermann, Paris.

Ruh, E.A.
  1971.   Minimal immersions of 2-spheres in S^4, Proc. Amer. Math. Soc., *28*, 219–222.

Ruh, E.A. and Vilms, J.
   1970. The tension field of the Gauss map, Trans. Amer. Math. Soc., *149*, 569–573.

Ryan, P.J.
   1969. Homogeneity and some curvature conditions for hypersurfaces, Tôhoku Math. J., *21*, 363–388.
   1971. Hypersurfaces with parallel Ricci tensor, Osaka J. Math., *8*, 251–259.

Sacksteder, R.
   1960. On hypersurfaces with no negative sectional curvatures, Amer. J. Math., *82*, 609–630.
   1962. The rigidity of hypersurfaces, J. Math. Mech., *11*, 929–939.

Santaló, L.A.
   1968. Curvatures absolutes totales de variedades contenidas en espcio eucliano, Acta Cient. Compos., *5*, 149–158.

Sard, A.
   1942. The measure of the critical points of differentiable maps, Bull. Amer. Math. Soc., *48*, 883–890.

Sasaki, S.
   1958. A global formulation of the fundamental theorem of the theory of surfaces in three dimensional euclidean space, Nagoya Math. J., *13*, 69–82.

Schouten, J.A.
   1921. Über die konforme Abbildung *n*-dimensionaler Mannigfaltigkeiten mit quadratischer Massbestimmung auf eine Mannigfaltigkeit mit euklidischer Massbestimmung, Math. Z., *11*, 58–88.
   1954. *Ricci-Calculus*, 2nd ed., Springer-Verlag, Berlin.

Schur, F.
   1886. Über den Zusammenhang der Räume konstanten Krümmungsmasses mit den projecktiven Räumen, Math. Ann., *27*, 609–613.

Shahin, J.K.
   1968. Some integral formulas for closed hypersurfaces in euclidean space, Proc. Amer. Math. Soc., *19*, 609–613.

Shiohama, K.
   1967. Cylinders in euclidean space E^{2+N}, Kōdai Math. Sem. Rep., *19*, 225–228.

Shiohama, K. and Takagi, R.
   1970. A characterization of a standard torus in E^3, J. Differential Geometry, *4*, 477–485.

Simon, U.
  1970. Probleme der lokalen und gloalen mehrdimensionalen Differential-geometrie, Manuscripta Math., *2*, 241–284.

Simons, J.
  1967. A note on minimal varieties, Bull. Amer. Math. Soc., *73*, 491–495.
  1968. Minimal varieties in Riemannian manifolds, Ann. of Math., *88*, 62–105.

Smyth, B.
  1967. Differential geometry of complex hypersurfaces, Ann. of Math., *85*, 246–266.
  1968. Homogeneous complex hypersurfaces, J. Math. Soc. Japan, *20*, 643–647.
  1973. Submanifolds of constant mean curvature, Math. Ann., *205*, 265–280.

Steenrod, N.
  1951. *Topology of Fibre Bundles*, Princeton Univ. Press, Princeton, New Jersey.

Sternberg, S.
  1964. *Lectures on Differential Geometry*, Prentice-Hall, Englewood Cliffs, New Jersey.

Stiel, E.
  1965. Isometric immersions of manifolds of nonnegative constant sectional curvature, Pacific J. Math., *15*, 1415–1419.
  1966. On immersions with singular second fundamental form operators, Proc. Amer. Math. Soc., *17*, 699–702.
  1967. Immersions into manifolds of constant negative curvature, Proc. Amer. Math. Soc., *18*, 713–715.

Stoker, J.J.
  1969. *Differential Geometry*, Wiley-Interscience, New York.

Szczarba, R.H.
  1969. On existence and rigidity of isometric immersions, Bull. Amer. Math. Soc., *75*, 783–787; Addendum, *76* (1970), 425.

Tai, S.S.
  1968. On minimum imbeddings of compact symmetric spaces of rank one, J. Differential Geometry, *2*, 55–66.

Takahashi, T.
  1966. Minimal immersions of Riemannian manifolds, J. Math. Soc. Japan, *18*, 380–385.
  1970. Homogeneous hypersurfaces in spaces of constant curvature, J. Math. Soc. Japan, *22*, 395–410.

1971. An isometric immersion of a homogeneous Riemannian manifold of dimension 3 in the hyperbolic space, J. Math. Soc. Japan, *23*, 649–661.

Takeuchi, M. and Kobayashi, S.
1968. Minimal imbeddings of *R*-spaces, J. Differential Geometry, *2*, 203–215.

Tanno, S.
1972. 2-dimensional complex in complex submanifolds immersed in complex projective spaces, Tôhoku Math. J., *24*, 71–78.

Teng, T.H. and Chen, B.-Y.
1966. On the compact orientable Riemannian manifolds immersed in euclidean space, Formosan Sci., *20*, 69–76.
1967. On the α-th curvatures of surfaces in euclidean spaces, Tamkang, J., *6*, 301–309.

Thomas, T.Y.
1936. On closed spaces of constant mean curvature, Amer. J. Math., *58*, 702–704; *59* (1937), 793–794.

Thorpe, J.A.
1964. Sectional curvatures and characteristic classes, Ann. of Math., *80*, 429–443.

Tompkins, C.
1939. Isometric imbedding of flat manifolds in euclidean space, Duke Math. J., *5*, 58–61.
1941. A flat Klein bottle isometrically imbedded in euclidean 4-space, Bull. Amer. Math. Soc., *47*, 508.

Vilms, J.
1972. Submanifolds of euclidean space with parallel second fundamentle form, Proc. Amer. Math. Soc., *32*, 263–266.

Voss, K.
1956. Einige differentialgeometrische Kongruenzsätze für geschlossene Flächen und Hyperflächen, Math. Ann., *131*, 180–218.

Wegner, B.
1974a. Eine Charakterisierung von Produkten geschlossener sphärischer Kurven, Math. Nachr., *59*, 229–234.
1974b. Codazzi-Tensoren und Kennzeichnungen sphärischer Immersionen, J. Differential Geometry, *9*, 61–70.

Weinstein, A.
1970. Positively curved *n*-manifolds in R^{n+2}, J. Differential Geometry, *4*, 1–4.

Weyl, H.
  1918. Reine Infinitesimalgeometrie, Math. Z., *26*, 384–411.
  1921. Zur Infinitesimalgeometrie: Einordnung der projektiven und der konformen Auffassung, Göttingen Nachr., 99–112.
  1939. On the volume of tubes, Amer. J. Math., *61*, 461–472.

White, J.
  1973. A global invariant of conformal mapping in space, Proc. Amer. Math. Soc., *38*, 162–164.

Willmore, T.J.
  1959. *An Introduction to Differential Geometry*, Oxford Univ. Press, London and New York.
  1965. Note on embedded surfaces, An. Sti. Univ. "Al. I. Cuza," Iaşi, Şect. Ia Mat., *11B*, 493–496.
  1968a. Mean curvature of immersed surfaces, An. Sti. Univ. "Al. I. Cuza," Iaşi, Şect. Ia Mat., *14*, 99–103.
  1968b. Curvature of closed surfaces in E^3, Acta Cientifica Compostelana, Universidad de Santiage de Compostela, *V*, 7–9.
  1971a. Mean curvature of Riemannian immersions, J. London Math. Soc., *3*, 307–310.
  1971b. Tight immersions and total absolute curvature, Bull. London Math. Soc., *3*, 129–151.

Willmore, T.J. and Jhaveri, C.S.
  1972. An extension of a result of Bang-Yen Chen, Quart. J. Math., *23*, 319–323.

Willmore, T.J. and Saleemi, B.A.
  1966. The total absolute curvature of immersed manifolds, J. London Math. Soc., *41*, 153–160.

Wilson, J.P.
  1965. The total absolute curvature of an immersed manifold, J. London Math. Soc., *40*, 362–366.

Wolf, J.A.
  1966. Exotic metrics on immersed surfaces, Proc. Amer. Math. Soc., *17*, 871–877.
  1967. *Spaces of Constant Curvature*, McGraw-Hill, New York.
  1968. Surfaces of constant mean curvature, Proc. Amer. Math. Soc., *19*, 1103–1110.

Wu, H.
  1971. A structure theorem for complete noncompact hypersurfaces of nonnegative curvature, Bull. Amer. Math. Soc., *77*, 1070–1071.

Yano, K.
  1957. *The Theory of Lie Derivatives and its Applications*, North-Holland, Amsterdam.
  1965a. *Differential Geometry on Complex and Almost Complex Spaces*, Pergamon, New York.
  1965b. Closed hypersurfaces with constant mean curvature in a Riemannian manifold, J. Math. Soc. Japan, *17*, 333–340.
  1970. *Integral Formulas in Riemannian Geometry*, Marcel Dekker, New York.

Yano, K. and Bochner, S.
  1953. *Curvature and Betti Numbers*, Annals of Mathematical Studies, No. 32, Princeton Univ. Press, Princeton, New Jersey.

Yano, K. and Chen, B.-Y.
  1971a. Minimal submanifolds of a higher dimensional sphere, Tensor, *22*, 369–373.
  1971b. On the concurrent vector fields of immersed manifolds, Kōdai Math. Sem. Rep., *23*, 343–350.

Yano, K., Houh, C.S., and Chen, B.-Y.
  1973. Intrinsic characterization of certain conformally flat spaces, Kōdai Math. Sem. Rep., *25*, 357–361.

Yano, K. and Ishihara, S.
  1969. Pseudoumbilical submanifolds of codimension 2, Kōdai Math. Sem. Rep., *21*, 365–382.
  1971. Submanifolds with parallel mean curvature vector, J. Differential Geometry, *6*, 95–118.

Yano, K. and Tani, K.
  1969. Integral formulas for closed hypersurfaces, Kōdai Math. Sem. Rep., *21*, 335–349.

Yau, S.-T.
  1974. Submanifolds with constant mean curvature, I, Amer. J. Math., *96*, 346–366.

Author Index

Subject Index